Thiamin Pyrophosphate Biochemistry

Volume I
Fundamentals, Pyruvate Decarboxylase, and Transketolase

Editors

Alfred Schellenberger, Ph.D.
Department of Biochemistry
Martin Luther University
Halle, German Democratic Republic

Richard L. Schowen, Ph.D.
Department of Chemistry
University of Kansas
Lawrence, Kansas

CRC Press, Inc.
Boca Raton, Florida

0283-4923

CHEMISTRY

Library of Congress Cataloging-in-Publication Data

Thiamin pyrophosphate biochemistry.

   Based on a conference held in Oct. 1984 at Haus
Buchenberg in the Harz Mountains, near Wernigerode,
GDR and sponsored by the Martin Luther University in Halle
and the Biochemical Society of the GDR.
   Includes bibliographies and index.
   Contents: v. 1. Fundamentals, pyruvate decarboxylase,
and transketolase -- v. 2. The pyruvate dehydrogenase
complex and prospects for the future.
   1. Vitamin B1--Congresses. 2. Thiamin pyrophosphate--
Congresses. I. Schellenberger, Alfred. II. Schowen,
Richard L. III. Martin-Luther-Universität
Halle-Wittenberg. IV. Biochemische Gesellschaft der
Deutschen Demokratischen Republik. [DNLM: 1. Thiamine--
congresses. 2. Thiamine Pyrophosphate--congresses.
QU 135 T422 1984]
QP772.T5T48 1988   574.1'926   87-10367
ISBN 0-8493-4681-9 (set)
ISBN 0-8493-4682-7 (v. 1)
ISBN 0-8493-4683-5 (v. 2)

   Direct all inquiries to CRC Press, Inc., 2000 Corporate Blvd., N.W., Boca Raton, Florida, 33431.

International Standard Book Number 0-8493-4682-7 (v. 1)
International Standard Book Number 0-8493-4683-5 (v. 2)
International Standard Book Number 0-8493-4681-9 (set)

Library of Congress Card Number 87-10367
Printed in the United States

PREFACE

Thiamin pyrophosphate is a cofactor for a variety of enzymes that catalyze reactions involving the fission and formation of carbon-carbon bonds. Its importance extends from fundamental questions of enzyme mechanisms through metabolic regulation to clinical science.

In 1981, the New York Academy of Sciences held a conference entitled "Thiamin: Twenty Years of Progress," under the joint chairmanship of Professors Henry Z. Sable and Clark J. Gubler. From this meeting, there issued a fine volume bearing the same title as the conference (*Ann. N.Y. Acad. Sci.*, 378, 1-470, 1982). This volume, as any number of workers in the area can testify, has been deeply influential in the promotion of research in thiamin chemistry and biochemistry.

One of the editors of the present volume, Professor Alfred Schellenberger, felt that by 1984 the rapid progress of work in this field merited the organization of a second international conference. Accordingly, this took place in October 1984, at Haus Buchenberg in the Harz Mountains, near Wernigerode in the German Democratic Republic. The meeting was jointly sponsored by the Martin Luther University in Halle and the Biochemical Society of the German Democratic Republic. Scientists from the Americas, Asia, and Europe attended, not a few of them having been present 2 years earlier in New York. The papers read and discussed at this meeting have given rise to the two volumes we now introduce.

The two volumes are organized into a total of six parts, three parts in each volume. While the articles in each volume and the articles in each part form coherent wholes, there is, of course, overlap and arbitrariness in the classification. It is nevertheless pleasing to see the broad panorama of biochemistry that one can survey from the vantage point of thiamin.

The first volume contains chapters that deal with quite basic aspects of the chemistry and biochemistry, in particular the fundamental enzymology, of thiamin. Included are papers on chemical matters and variously phosphorylated thiamins and on two well-studied enzyme systems, pyruvate decarboxylase and transketolase.

Part I begins with the basic chemistry and biochemistry of thiamin and its derivatives, observed from many different points of view. Kluger and Gish address stereochemical problems connected with both chirality at the carbon center activated by thiamin and conformation of bonds about this center. More general conformational questions are considered theoretically by Friedemann, Richter, and Gründler and experimentally by Flatau and Zschunke. Dubois and El Hage Chahine and Zoltewicz, Uray, and Kriessmann present detailed results on chemical reactions of thiamin and related compounds. Finally, Ostrovsky examines metabolic aspects of variously phosphorylated forms of thiamin.

The contributions of Part II present results on pyruvate decarboxylase, an enzyme that has become a paradigm for problems of thiamin pyrophosphate biochemistry. Some chapters concern structure, some mechanism, and some both structure and mechanism.

The current status of the biochemistry of transketolase, another thiamin-pyrophosphate dependent enzyme of broad significance, is recounted in Part III by Kochetov and Gubler.

The second volume addresses aspects of the currently very active field of the pyruvate dehydrogenase enzyme complex, and also points the way toward future studies in the thiamin field.

Parts I and II of this volume contain papers on the pyruvate dehydrogenase enzyme complex, the importance of which is not only central to thiamin pyrophosphate biochemistry, but also for much more general questions of cellular metabolism. The papers that address structure are gathered in Part I. Part II consists of papers about the regulation of the complex.

In Part III of Volume II, we have introduced an unusual feature. There are many powerful techniques of biochemistry, particularly of molecular enzymology, which have either not been applied to thiamin systems or have not, in the opinion of the editors, been applied with sufficient frequency. We therefore asked several experts in these areas to write about their techniques in order to encourage their application in the thiamin area. Eklund has written on X-ray crystallography, Damerau on spin-labeling studies, Christen and Kirsten on catalysis in crystalline systems, and Christen, Lubini, and Cogoli on paracatalytic inactivation. Fischer, Mech, and Bang have contributed a brief chapter on the important topic of the initial attainment of the three-dimensional structure of enzymes after their synthesis. In each case, reference is made to applications to thiamin systems if they have been studied, but extensive examples of other applications are also treated.

We hope and believe that readers will find these two volumes useful in assessing the current state of affairs in thiamin pyrophosphate biochemistry. If so, we owe this success to our colleagues who wrote effective chapters. The editors offer them cordial thanks.

THE EDITORS

Alfred Schellenberger, Ph.D., is Professor and Head of the Department of Biochemistry at the Martin Luther University of Halle-Wittenberg in the German Democratic Republic. He received his doctoral training at the same university, completing the degree in 1956 under the direction of Professor W. Langenbeck.

In addition to having published over 130 original papers and reviews, he holds 18 patents in the area of enzyme technology. In 1970 he became a member of the German Academy of Natural Scientists, the Leopoldina, and has served in its Senate since 1985. He is a member of the Council of the Biochemical Society of the German Democratic Republic. In 1974 and 1982, he received the award "Forschungspreis I" of his university.

Professor Schellenberger's major research interests are the chemistry of 2-oxo acids, the catalytic mechanisms of thiamin-pyrophosphate enzymes, and mechanistic aspects of the activation of pyruvate decarboxylase.

Richard L. Schowen, Ph.D., is Solon E. Summerfield Professor of Chemistry and Biochemistry at the University of Kansas. His doctoral degree was earned under Professor C. G. Swain at the Massachusetts Institute of Technology in 1962.

Professor Schowen's research is in the area of mechanisms of organic and biochemical reactions, origins of the catalytic power of enzymes, and the application of isotopic methods to problems in these areas. He has published approximately 100 papers on these subjects.

He is a member of the American Chemical Society, the American Society of Biological Chemists, and is a Fellow of the American Association for the Advancement of Science. He serves on the Editorial Board of the *Journal of the American Chemical Society* and is a recipient of the Dolph Simons, Sr. Award for Research Achievement in Biomedical Science.

CONTRIBUTORS

Olufemi B. Akinyosoye, Ph.D.
Instructor
Department of Chemistry
Passaic County College
Paterson, New Jersey

Francisco J. Alvarez, Ph.D.
Research Associate
Department of Pharmaceutical Chemistry
University of Kansas
Lawrence, Kansas

George Dikdan, M.S.
Department of Chemistry
Rutgers University
Newark, New Jersey

Jacques-Emile Dubois, D.Sc.
Professor
Department of Physical Chemistry
Universite de Paris VII
Paris, France

Jean-Michel El Hage Chahine, D.Sc.
Chargé de Recherche
CNRS
Université Paris
Paris, France

Sabine Flatau, Dr. rer. nat.
Assistant
Department of Biochemistry
Martin Luther University
Halle, G.D.R.

Rudolf Friedemann, Dr. sc. nat.
Department of Chemistry
Martin Luther University
Halle, G.D.R.

Gerald Gish, Ph.D.
Research Fellow
Department of Chemistry
Max Planck Institute for Experimental
  Medicine
Göttingen, F.R.G.

Wolfgang Gründler, Dr. sc. nat.
Professor
Department of Chemistry
Humbolt University of Berlin
Berlin, G.D.R.

Clark J. Gubler, Ph.D.
Professor
Department of Biochemistry
Kuwait University
Kuwait, Kuwait
Professor Emeritus
Brigham Young University
Provo, Utah

Gerhard Hübner, Dr. sc. nat.
Department of Biotechnology
Martin Luther University
Halle, G.D.R.

Frank Jordan, Ph.D.
Professor and Chairman
Department of Chemistry
Rutgers University
Newark, New Jersey

Ronald Kluger, Ph.D.
Professor
Department of Chemistry
University of Toronto
Toronto, Ontario, Canada

G. A. Kochetov
A.N. Belozersky Laboratory of Molecular
  Biology and Bioorganic Chemistry
Moscow State University
Moscow, U.S.S.R.

Stephan König, Dr. rer. nat.
Assistant
Department of Biochemistry
University of Halle
Halle, G.D.R.

Ingo Kriessmann, Ph.D.
Department of Organic Chemistry
University of Graz
Graz, Austria

Zbigniew H. Kudzin, Ph.D.
Assistant Professor
Department of Chemistry
Institute of Chemistry
Narutowicza, Lodz, Poland

Donald J. Kuo, Ph.D.
Research Associate
Institute for Cancer Research, Fox Chase
Philadelphia, Pennsylvania

Yu.M. Ostrovsky
Professor and Director
Institute of Biochemistry
U.S.S.R. Academy of Sciences
Grodno, U.S.S.R.

Dieter R. Petzold, Dr. rer. nat.
Department of Molecular Biology
Institute of Virology and Epidemiology
Berlin, G.D.R.

Mohammed Rahmatullah
Department of Biochemistry
Kansas State University
Manhattan, Kansas

Donald Richter
Department of Chemistry
Martin Luther University
Halle, G.D.R.

A. Schellenberger
Professor
Sektion Biowissenschaften
Wissenschaftsbereich Biochemie
Martin Luther Universität
Halle, G.D.R.

Richard L. Schowen, Ph.D.
Summerfield Professor of Chemistry and
  Biochemistry
Department of Chemistry
University of Kansas
Lawrence, Kansas

Manfred Sieber, Dr. rer. nat
Institut für Physiologische Chemie
Universität des Saarlandes
Homburg/Saar, F.R.G.

Johannes Ullrich, Dr. rer. nat.
Professor
Biochemisches Institut
Universität Freiburg
Freiburg, F.R.G.

Georg Uray, Ph.D.
Associate Professor
Department of Organic Chemistry
Karl-Franzens University
Graz, Austria

Hartmut Zehender, Dr. rer. nat.
Klinisch-Chemisches Institut
Klinikum der Stadt Mannheim
Mannheim, F.R.G.

John A. Zoltewicz, Ph.D.
Professor
Department of Chemistry
University of Florida
Gainesville, Florida

TABLE OF CONTENTS

Volume I

PART I. THE FUNDAMENTAL CHEMISTRY AND BIOCHEMISTRY OF THIAMIN AND ITS DERIVATIVES

PART II. PYRUVATE DECARBOXYLASE: STRUCTURE AND MECHANISM OF ACTION

Volume II

PART I. THE PYRUVATE DEHYDROGENASE ENZYME COMPLEX: STRUCTURE

Part I
The Fundamental Chemistry and Biochemistry of Thiamin and Its Derivatives

Chapter 1

STEREOCHEMICAL ASPECTS OF THIAMIN CATALYSIS

Ronald Kluger and Gerald Gish

TABLE OF CONTENTS

I. INTRODUCTION

The initial covalent intermediate in the thiamin-catalyzed decarboxylation of pyruvate is 2-[(2-hydroxy-2-carboxy)-ethyl] thiamin which we refer to as lactylthiamin.[1-3] The transformation of pyruvate into lactylthiamin provides the "Umpolung" of pyruvate's reactivity necessary for decarboxylation and formation of hydroxyethylthiamin, 2-[(2-hydroxy)-ethyl] thiamin, the precursor of acetaldehyde. The corresponding diphosphates are the covalent intermediates in thiamin-diphosphate-dependent enzymic reactions of pyruvate.[2,4]

Stereochemistry is an intimate part of these transformations, and insight into the mechanism of the catalytic transformations can be obtained from analysis of the stereochemistry. This chapter deals with two of the stereochemical aspects of this system: (1) the relation between conformation and reactivity of lactylthiamin and (2) the resolution of chiral intermediates.

II. CONFORMATION AND REACTIVITY OF LACTYLTHIAMIN

During a catalytic cycle, the only covalent structural changes associated with thiamin occur at the C-2 position of the thiazolium ring. A proton is lost from this position to form the ylid which adds to the carbonyl group of pyruvate, generating lactylthiamin. After decarboxylation and protonation, the hydroxyethyl derivative is produced. This undergoes elimination to produce acetaldehyde while the ylid is regenerated and reprotonated. The elimination process which generates acetaldehyde should be mechanistically similar to the reverse of the addition process which forms lactylthiamin. What relationship is there between the accessible and favored conformations of the intermediates derived from thiamin and their reactivity patterns?

A. Controlling Factors

There are steric and electronic constraints on conformations in the addition of nucleophiles to carbonyl groups.[5,6] The addition of the thiazolium-derived ylid of thiamin to the carbonyl group of pyruvate will be subject to these constraints. First, we consider the thiazolium ylid as a point-charge nucleophile, and analyze variations in the angle of attack of that pair of electrons upon the plane of the carbonyl group of pyruvate. This geometry in other carbonyl addition reactions has been the subject of extensive studies that stem from the observations of Buergi et al.[6] The approach of the nucleophile is electronically most favorable from an anticlinal alignment relative to the carbonyl group and in a plane perpendicular to that of the carbonyl group (Figure 1).

This angle of approach establishes an axis between the carbonyl carbon atom and the nucleophile. The thiazolium ylid contains its localized negative charge as a pair of electrons in a p orbital in the plane of the ring. The axis we have defined accommodates complete rotation of the plane of the thiazolium ring. Any further restraints on conformation are steric and electrostatic.[6] In the case of an enzymic reaction, the geometry of binding sites define the relationship. The bulky methylene-pyrimidine substituent on C-3 of the thiazolium ring provides strong steric direction. Based on the X-ray structure of phosphalactylthiamin,[7] we see that none of the three substituents at the C2-alpha carbon of lactylthiamin (carboxyl, methyl, hydroxyl) can be in the plane of the thiazolium ring since they will collide with the methylene-pyrimidine ring. To avoid this steric interaction, a substituent is best held out of the plane of the thiazolium ring. If the single interaction between the substituent and methylene pyrimidine is large, then the lowest energy conformation will have the substituent perpendicular to the plane of the ring (Figure 2, conformer A). If the other substituents also have significant interactions, then an arrangement where the substituents straddle the ring plane is best (Figure 2, conformer B). Since the thiazolium ring is not symmetrical in its plane, each conformation will have an isomeric form in which the thiazolium ring is rotated by 180°.

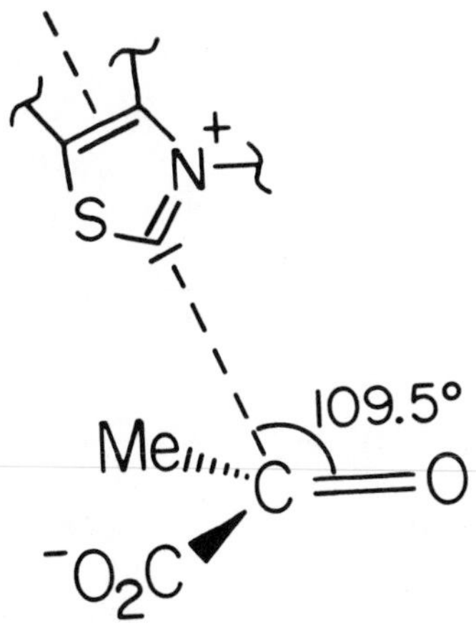

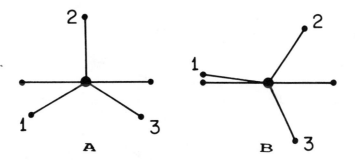

FIGURE 1. The axis defining the best angle for approach to the carbonyl group of pyruvate.

FIGURE 2. Possible arrays of substituents (1, 2, 3) on the C2-alpha carbon relative to the plane of the thiazolium ring. Conformation A should result if one group is large compared to the other two, and conformation B should result if the two bulkiest groups are comparable in size.

For intermediate cases, some combination of the two limits should result. The angle of the largest substituent relative to the plane of thiazolium, by this analysis, will be between 90 and 60°. In the case of phosphalactylthiamin, the ethyl phosphonate group is large compared to the hydroxyl and methyl groups at C2-alpha.[7] The resulting structure has a 90° angle between the plane of the thiazolium ring and the ethyl phosphonate group as in conformer A of Figure 2.

We can extrapolate to lactylthiamin and propose that the carboxylate will occupy a position similar to that of the phosphonate in the analog (substituent 2 in Figure 2, conformer A). Then, by application of the Hammond postulate,[8] we propose that the conformation will also hold for the transition state, which leads to the formation of this intermediate.

B. Electrostatic Selectivity

The perpendicular arrangement of the two planes still permits two possible relationships (Figure 3). We define these in terms of the relationship of the carbonyl group of pyruvate and the C2-N3 bond of the thiazolium ring. The nitrogen can be synperiplanar or antiperiplanar with respect to the carbonyl oxygen. The choice between conformations can probably be settled in terms of electrostatic interactions if localized charges can be determined. The

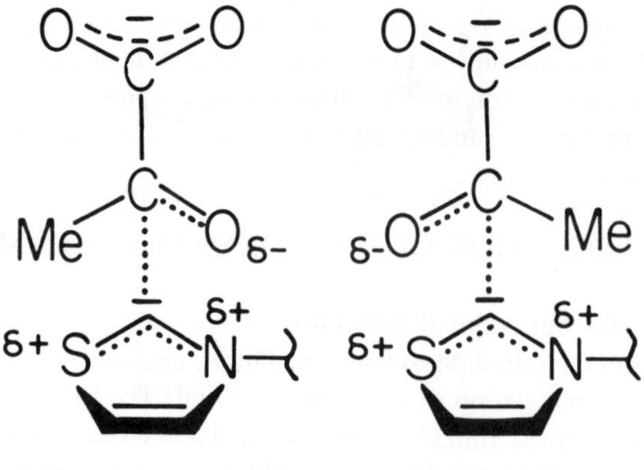

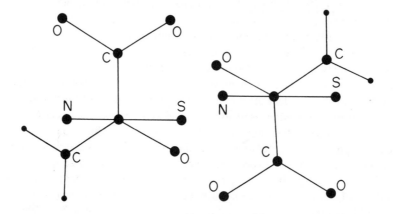

FIGURE 3.   Two likely arrangements for addition of the thiamin ylid to pyruvate which maximize electrostatic stabilization.

FIGURE 4.   The conformations of lactylthiamin which are likely to be favored for its decarboxylation. The enantiomers have been chosen arbitrarily.

carbonyl group acquires negative charge in the transition state for addition, and the closest positioning of the positive charge of the thiazolium ring should provide maximum electrostatic stabilization. The thiazolium ring is usually drawn with the positive charge localized on nitrogen, but an important resonance structure will have a positive charge on sulfur and no charge on nitrogen.[9-11]

C. Stereoelectronic Restrictions on Decarboxylation

The most favorable conformation for decarboxylation of lactylthiamin should be decided by stereoelectronic factors. The loss of carbon dioxide from lactylthiamin leaves a pair of electrons in an orbital derived from the sigma bond that connected the carboxylate group to C-2 of the lactyl group. The incipient anion is stabilized by overlap with the pi system of the thiazolium ring. In order for overlap to occur, the orbital must be in a plane perpendicular to the plane of the thiazolium ring, leading to two possible conformers (Figure 4). These conformers are the same as two of the four possibilities which are favored for the formation of lactylthiamin (Figures 2 and 3), assuming that motion to the transition state will be minimized.[12]

Thus, the conformation of the transition state which leads to decarboxylation of lactyl-thiamin and that which leads to its formation correlate with nearly the same conformations of this intermediate. Therefore, the intermediate will not have to undergo any significant movement to proceed from one step of the catalytic process to the next. One can therefore apply the principle of least motion[12] to this system with confidence. An important corollary is that an enzymic process involving lactylthiamin diphosphate will not require special direction or the expenditure of binding energy to induce conformational changes in the coenzyme or its adducts.

III. RESOLUTION OF HYDROXYETHYLTHIAMIN

A. Chiral Intermediates but Achiral Reactants

Although thiamin and thiamin diphosphate are achiral, the covalent intermediates formed during the catalytic decarboxylation of pyruvate are chiral. The first adduct, lactylthiamin, contains a carbon atom, derived from C-2 of pyruvate, at a stereogenic center.[13] This material undergoes decarboxylation and protonation to yield hydroxyethylthiamin, which is also chiral and stereogenic at the same center.

The intermediates in the reaction catalyzed by pyruvate decarboxylase, lactylthiamin diphosphate, and hydroxyethylthiamin diphosphate are both chiral, but the absolute stereochemistry of neither is known for any enzymic process. Since the adducts are not released as products of the reaction, but function as bound intermediates, the intermediates are present only at a low steady-state level. This is in marked contrast to systems in which substrates and products are analyzed. We decided to attack the problem by preparing resolved materials and monitoring their interactions with enzymes that have been freed of thiamin diphosphate. We expect the enzyme to accept the chiral adduct of proper stereochemistry and to continue the catalytic process, converting itself to the holoenzyme. This is not necessarily the case, however. The apoenzyme does not normally combine with adducts of the coenzyme but rather the coenzyme itself, generating the adducts from bound substrates. In an ordered binding system, we would not expect the adducts to bind if added exogenously. Another possibility is that stereospecificity will not be manifested for exogenously added adducts. One or both enantiomers may bind; one or both may react. If an adduct binds but does not react, it should function as an inhibitor toward activation of the apoenzyme by thiamin diphosphate. Still, if the adducts do bind and activate stereospecifically, then knowledge of the absolute stereochemical relationship of the two adducts will limit the number of possible mechanisms.

B. Isolation of Chiral Intermediates

The adduct of radioactively labeled acetaldehyde and thiamin diphosphate, hydroxyethylthiamin diphosphate, was prepared by Krampitz and co-workers[14] who demonstrated that it reacts with the apoenzyme of wheat germ pyruvate decarboxylase, releasing radioactive acetaldehyde in a burst followed by a slow turnover as thiamin diphosphate dissociates from the enzyme. Holzer and Beaucamp[15] reported isolating both the acetaldehyde adduct and the pyruvate adduct. However, the adduct which is supposed to be derived from pyruvate cannot be lactylthiamin diphosphate since it has been shown that the material is decomposed in methanol,[2] and the reported isolation procedure involved heating the material in methanol.[15] Thus, the pyruvate adduct has not yet been isolated from an enzyme, although we have developed a chemical synthesis of racemic lactylthiamin diphosphate.[4]

The previous work on chiral intermediates in the pyruvate decarboxylase system is limited. Most notably, Ullrich and Mannschreck[16] isolated optically active hydroxyethylthiamin diphosphate from a large-scale reaction of pyruvate dehydrogenase. We compare their results with those obtained by chemical resolution.[17] There is also a report on the use of resolved hydroxyethylthiamin in nutrition studies, but the report does not reveal the resolving procedure, nor are the specific properties of the resolved material mentioned.[18]

C. Preparation of Chiral Intermediates: Hydroxyethylthiamin

We have developed an improved synthesis of hydroxyethylthiamin, a procedure for resolution of hydroxyethylthiamin and determination of its absolute stereochemistry, and a
method for converting hydroxyethylthiamin to the enzymically active material, hydroxyethylthiamin diphosphate.[17] Enzyme kinetic studies reveal that both enantiomers may bind
to the enzyme and may function as inhibitors with respect to thiamin diphosphate as activators
of the apoenzyme.[19]

Racemic hydroxyethylthiamin was prepared by the reaction of thiamin chloride in ethanol
with acetaldehyde at $-5°C$ in the presence of two equivalents of sodium ethoxide. After
reaction was complete, the mixture was filtered and the precipitated hydroxyethylthiamin
was collected and washed.

1. Resolution of Hydroxyethylthiamin

A 1:1 adduct of (+)-2,3-dibenzoyl-D-tartaric acid and hydroxyethylthiamin was formed
in alkaline ethanol. The solution was evaporated to dryness and the residue was recrystallized
slowly from water. The crystals were collected and washed. They were recrystallized until
a constant measured optical rotation was obtained. The specific rotation at the sodium D
line of resolved hydroxyethylthiamin was determined to be $-12.5°$ for a solution in water
of 0.7% concentration at 22°C. The isolated material had optical rotation properties similar
to the diphosphate isolated by Ullrich and Mannschreck from pyruvate dehydrogenase which
had a specific activity of $-10 \pm 2°$. Shiobara et al.[18] report specific rotations of -13.5
and $+12.5°$ for material they used in nutritional experiments.

We used circular dichroism spectroscopy to measure optical activity in the region in which
the molecule absorbs light. The resolved material gives two peaks which coincide with the
absorbances of the thiazolium and pyrimidine chromophores. This also establishes that the
resolving agent has been removed.

2. Absolute Configuration of Resolved (+) Hydroxyethylthiamin

It is essential to know the relative configurations of reactants and products in this series.
In order to do so, we need to know the absolute stereochemistry associated with the optical
properties we measure. We first tried to convert the resolved material to compounds of
known absolute configuration. We attempted to develop an oxidative degradation procedure
of hydroxyethylthiamin which could yield lactic acid whose stereocenter is derived from the
C2-alpha carbon of hydroxyethylthiamin. Unfortunately, pure material could not be produced
in our attempts. Fortunately, in collaboration with Dr. George DeTitta (Medical Foundation
of Buffalo, N.Y.), the absolute configuration of (+)-hydroxyethylthiamin has been determined indirectly by X-ray crystallography. The crystals of the dibenzoyltartrate adduct are
needles which are not suitable for analysis. Therefore, we converted (+)-hydroxyethyl-
thiamin to a related chiral material without disturbing the stereocenter. The absolute configuration and details of the conversion procedure are reported in detail elsewhere.

3. Pyrophosphorylation of Hydroxyethylthiamin

As stated earlier, we resolved hydroxyethylthiamin in order to investigate the stereospecificity of the enzymes which utilize its diphosphate derivative. Therefore, hydroxyethylthiamin must be pyrophosphorylated on the primary hydroxyl of the C-5 side chain, while
the secondary hydroxyl group of the hydroxyethyl side chain must remain free. Selective
pyrophosphorylation was accomplished by using the procedure we developed for the pyrophosphorylation of thiamin thiazolone.[19] The reaction must be carefully controlled since
reaction at the secondary hydroxyl does occur under more vigorous conditions.[20] The material
thus obtained was purified by ion-exchange chromatography. A comparison sample of racemic hydroxyethylthiamin diphosphate was prepared by condensation of acetaldehyde with

thiamin diphosphate and shown to have similar activity toward the apoenzyme of pyruvate decarboxylase. The details of the resolution have been reported,[17] and further reports on other aspects of this review are planned.

ACKNOWLEDGMENTS

We thank the Natural Sciences and Engineering Research Council of Canada for continued operating grant and fellowship support. The group in Toronto whose work we have discussed also included David Pike, Glenn Kauffman, Victoria Stergiopoulos, and Khashayar Karimian. R. K. thanks Professor Alfred Schellenberger and the organizers of the conference in Wernigerode for their hospitality and enthusiasm.

REFERENCES

1. **Breslow, R.,** The mechanism of thiamine action. IV. Evidence from studies on model systems, *J. Am. Chem. Soc.,* 80, 3719, 1958.
2. **Kluger, R., Chin, J., and Smyth, T.,** Thiamin-catalyzed decarboxylation of pyruvate. Synthesis and reactivity analysis of the central, elusive intermediate, alpha-lactylthiamin, *J. Am. Chem. Soc.,* 103, 884, 1981.
3. **Krampitz, L. O.,** *Thiamin Diphosphate and Its Catalytic Functions,* Marcel Dekker, New York, 1970.
4. **Kluger, R. and Smyth, T.,** Interaction of pyruvate-thiamin diphosphate adducts with pyruvate decarboxylase. Catalysis through "closed" transition states, *J. Am. Chem. Soc.,* 103, 1214, 1981.
5. **Wipke, W. T. and Gund, P.,** Simulation and evaluation of chemical synthesis. Congestion: a conformation-dependent function of steric environment at a reaction center. Application with torsional terms to stereoselectivity of nucleophilic additions to ketones, *J. Am. Chem. Soc.,* 98, 8107, 1976.
6. **Buergi, H. B., Dunitz, J. D., Lehn, J. M., and Wipff, G.,** Stereochemistry of reaction paths at carbonyl centers, *Tetrahedron,* 30, 1563, 1974.
7. **Turano, A., Furey, W., Pletcher, J., Sax, M., Pike, D., and Kluger, R.,** Synthesis and crystal structure of an analogue of 2-(alpha-lactyl)thiamin, racemic methyl 2-hydroxy-2-(2-thiamain)ethylphosphonate chloride trihydrate. A conformation for a least-motion, maximum-overlap mechanism for thiamin catalysis, *J. Am. Chem. Soc.,* 104, 3089, 1982.
8. **Hammond, G. S.,** A correlation of reaction rates, *J. Am. Chem. Soc.,* 77, 334, 1955.
9. **Haake, P. and Miller, W. B.,** A comparison of thiazoles and oxazoles, *J. Am. Chem. Soc.,* 85, 4044, 1963.
10. **Hogg, J. L.,** Consideration of a new catalytic role for the thiazolium sulfur atom of the coenzyme thiamin pyrophosphate, *Bioorg. Chem.,* 10, 233, 1981.
11. **Jordan, F.,** Theoretical calculations on thiamin and related compounds. II. Conformational analysis and electronic properties of 2-(alpha-hydroxyethyl)thiamin, *J. Am. Chem. Soc.,* 96, 808, 1976.
12. **Hine, J.,** The principle of least nuclear motion, *Adv. Phys. Org. Chem.,* 15, 1, 1977.
13. **Siegel, J. and Mislow, K.,** Stereoisomerism and local chirality, *J. Am. Chem. Soc.,* 106, 3319, 1984.
14. **Krampitz, L. O., Suzuki, I., and Greull, G.,** The role of thiamin diphosphate in catalysis, *Fed. Proc.,* 20, 974, 1961.
15. **Holzer, H. and Beaucamp, K.,** Nachweis und Charakterisierung von Alpha-Lactyl-Thiaminpyrophosphat ("Aktives Pyruvat") und Alpha-Hydroxyaethyl-Thiaminpyrophosphat ("Akitver Acetaldehyd") als Zwischenprodukte der Decarboxylierung von Pyruvat mit Pyruvatdecarboxylase aus Bierhefe, *Biochim. Biophys. Acta,* 46, 225, 1961.
16. **Ullrich, J. and Mannschreck, A.,** Studies on the properties of $(-)$-2-alpha-hydroxyethyl-thiamine pyrophosphate ("active acetaldehyde"), *Eur. J. Biochem.,* 1, 110, 1967.
17. **Kluger, R., Stergiopoulos, V., Gish, G., and Karimian, K.,** Chiral intermediates in thiamin catalysis: resolution and pyrophosphorylation of hydroxyethylthiamin, *Bioorg. Chem.,* 13, 227, 1985.
18. **Shiobara, Y., Sato, N., Yogi, K., and Murakami, M.,** Studies on hydroxyethylthiamine and related compounds. III. Nature of thiamine-active compounds in the liver following oral administration of hydroxyalkylthiamine-deficient rates, *J. Vitaminol.,* 11, 302, 1965.
19. **Kluger, R., Gish, G., and Kauffman, G.,** Interaction of thiamin diphosphate and thiamin thiazolone diphosphate with wheat germ pyruvate decarboxylase, *J. Biol. Chem.,* 259, 8960, 1984.
20. **Gish, G.,** Ph.D. thesis, University of Toronto, Toronto, Canada, 1985.

Chapter 2

QUANTUM-CHEMICAL STUDIES IN THE THIAMIN SYSTEM

Rudolf Friedemann, Donald Richter, and Wolfgang Gründler

Information about the structure and the conformational properties of thiamin and its analogs was obtained mainly from crystallographic data,[1-3] PMR and CMR studies,[4-7] and theoretical investigations.[8-10] Until now, the conformational behavior and the electronic structure of the isolated thiamin molecule, and especially of its pyrophosphate ester (TPP), were investigated by semiempirical quantum chemical methods and empirical atomic potential models involving a rotation of the two aromatic rings with respect to the bridge methylene group only.[8,9] Therefore, we have performed conformational calculations on thiamin cation (TH$^+$) and thiamin pyrophosphate (TPP), including variation of torsion angles of the rings and those of the side chain at the C5 atom of the thiazolium ring.

Moreover, the conformational studies were extended to the ylid-like structures of thiamin and TPP formed by the loss of a proton from the C2 atom of the thiazolium ring. These structures are important in the enzymatic mechanism proposed. In order to determine the preferred positions for binding of divalent cations to thiamin systems as well as to investigate their influence on conformational changes of these molecules, the interaction of Mg^{2+} ions with thiamin ylid (TH), TPP, and TPP ylid (TPP$^-$) was calculated.

For our studies, we have used the model of local electron pairs (LEP) — a model potential allowing the treatment of intra- and intermolecular interactions of large molecules.[11,12] Within the LEP model, the molecular system is built up from potential centers for the atomic cores and the valence electrons. Each electron pair is described by a spherical gaussian function.

$$\phi_i(r) = (2/\pi\rho_i^2) \exp[-(r - R_i)^2/\rho_i^2]$$

where R_i is the position vector of the gaussian function, and the orbital radius ρ_i characterizes the size of the electron pair. These are the only parameters of the model. For calculating the total interaction energy ΔE_G in addition to the electrostatic part ΔE_{Coul} the repulsion ΔE_{Rep} and dispersion contributions ΔE_{Disp} are taken into account:

$$\Delta E_G = \Delta E_{Coul} + \Delta E_{Rep} + \Delta E_{Disp}$$

In some ways, the LEP model is similar to the EPEN/2 method developed by Scheraga et al.[13] For our conformational studies in thiamin and TPP, the torsion angles defined in Figure 1 are considered. The relative orientation of the thiazolium and pyrimidine rings are indicated by the torsion angles φ_T and φ_P. Their notation corresponds to that defined by Pletcher and Sax[1] with the usual convention of the torsion angles. $\varphi_T = \varphi_P = 0°$ corresponds to the coplanar arrangement of the two rings. The orientation of the side chains is characterized by three and five torsion angles, respectively, given in Figure 1. The pyrophosphate group is assumed to have the stable staggered orientation. The geometry used in the computations was taken from crystallographic X-ray data of thiamin chloride (THC)[2] and TPP.[1] For characterizing the conformers of thiamin and TPP we have used the three basic conformations S ($\varphi_T = \pm100°$, $\varphi_P = \pm150°$), F ($\varphi_T = 0°$, $\varphi_P = \pm90°$) and V ($\varphi_T = \pm90°$, $\varphi_P = \pm90°$) observed in crystal structure analysis.[3]

First, we calculated the conformational energy maps with respect to the ring torsion angles. Figure 2 presents one of these for the thiamin cation. The total interaction energy was computed for each 20° rotation of the angles φ_T and φ_P, and the two-dimensional energy surface is plotted indicating energies in kilojoules per mole above the lowest conformation.

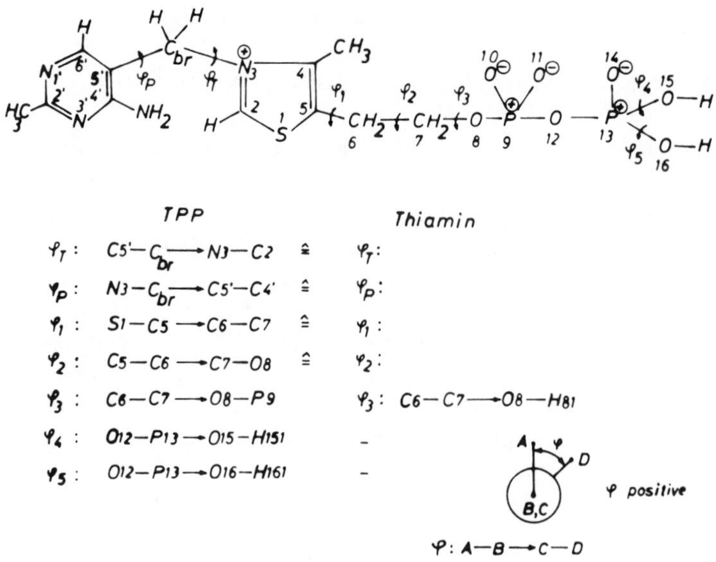

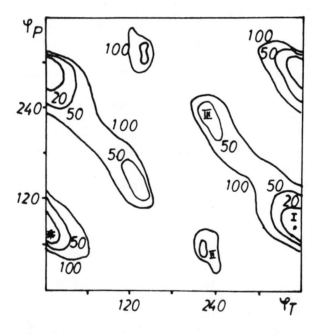

FIGURE 1. Torsion angles of TPP and thiamin taken into account.

FIGURE 2. Conformational map of thiamin cation (isoenergy lines in kJ/mol). $E_I = 0$; $E_{II} = 29$; $E_{III} = 34$; $E_* = 49$ (crystal structure).

The favored angle combinations (φ_T, φ_P) and the relative energies are summarized in Figure 2. The crystallographic geometry of THC is in a preferred region of the map. Similar results are obtained for the conformational maps of thiamin ylid and TPP. These findings essentially confirm the results from Jordan[8] within an empirical potential model. However, in the LEP model, the conformational barriers are higher and therefore the preferred regions smaller. Starting from the energetically favored combinations of the angles φ_T and φ_P a complete optimization of all torsion angles defined in Figure 1 was performed. The results for the thiamin cation and thiamin ylid are shown in Table 1. In both systems the conformational

Table 1
TORSION ANGLES (IN DEGREES) AND RELATIVE ENERGIES (kJ/mol) OF THE PREFERRED CONFORMATIONS OF THIAMIN CATION AND THIAMIN YLID

Conformation	φ_T	φ_P	φ_1	φ_2	φ_3	E_r
Thiamin cation						
(1)	−6.7	70.4	−5.2	60.0	7.5	0
(2)	−6.6	70.4	1.6	−58.4	−5.7	2.3
(3)	−6.3	70.3	−21.1	66.7	171.0	8.0
Crystal structure	2.6	76.8	−66.8	−64.6	98.7	48.7
Thiamin ylid						
(1)	121.8	−23.8	−104.4	−36.3	25.8	0
(2)	119.4	−24.2	73.8	−0.4	6.0	17.2
(3)	119.9	−24.8	−34.7	−38.7	18.2	19.1

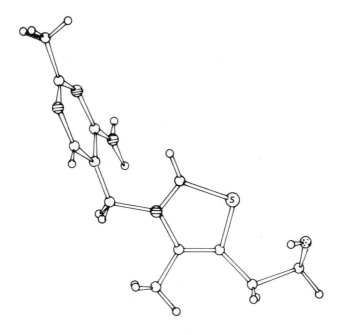

FIGURE 3.   Thiamin cation (1).

energy is essentially determined by the angles φ_T and φ_P and we find a flexibility of the side chain. The preferred conformations of the thiamin systems are presented in Figures 3 to 5. In the most favored conformation of thiamin cation we find an F-like form of the two rings. The OH group of the side chain is directed to the S1–C5 bond of the thiazolium ring. The crystal structure shows nearly the same arrangement of the rings. In comparison with thiamin cation (1), the side chain is more folded, but we find a similar interaction of the OH group with the thiazolium ring. The conformation of thiamin cation (3) (Table 1) with an extended side chain is less stable by only 8 kJ/mol. A comparison of the stable confor-mations of thiamin cation and thiamin ylid shows that the loss of the proton from the C2 atom causes significant conformational changes with respect to the orientation of the rings, while the characteristic back-folded structure of the side chain is maintained in the ylid. The

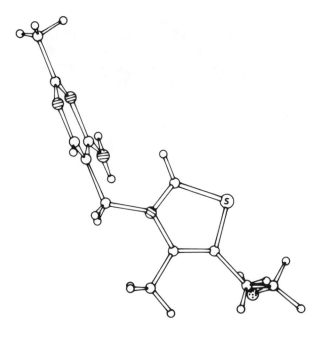

FIGURE 4.   Thiamin (crystal structure).

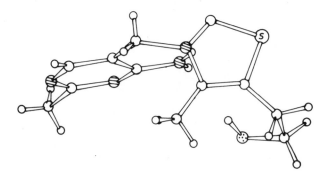

FIGURE 5.   Thiamin ylid (1).

interaction of a Mg^{2+} ion with thiamin ylid is favored at the thiazolium ring (Figure 6). In the most stable structure the Mg^{2+} ion is positioned 2.0 Å above the center of the thiazolium ring. The arrangement of the side chain enables an interaction of the Mg^{2+} ion with the oxygen atom (R_{Mg-08} = 2.8 Å). The interaction energy of the relaxed complex is −1043 kJ/mol. The structure of the Mg^{2+}-thiamin ylid is significantly different from that of the isolated ylid molecule in both the orientation of the rings and the side chain. The attack of a Mg^{2+} ion to the center of the pyrimidine ring is less favored and results in an interaction energy of only −861 kJ/mol.

Some years ago, Scrocco and Tomasi[14] and Pullman[15] have shown that the molecular electrostatic potential (MEP) and the molecular electrostatic field (MEF) are useful tools for obtaining qualitative information about the reactive behavior of large molecules. Within the LEP model, the MEP $V(r)$ and the MEF $E(r)$ on the position r can be obtained for biomolecules in a relatively simple way.[16] The MEP $V(r)$ is proportional to the electrostatic interaction energy between the charge distribution of the molecule and a proton localized on the position r. In order to have a suitable reference plane for calculating the MEP we

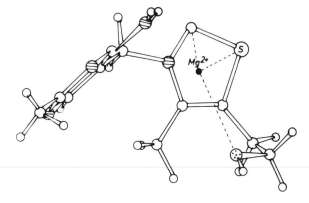

FIGURE 6. Mg^{2+}-thiamin ylid (1).

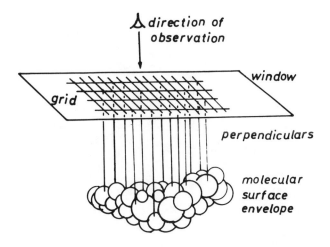

FIGURE 7.   Representation of a surface potential calculation.

have used the procedure proposed by Lavery and Pullman.[17] Its essence consists of calculating the potentials on the surface envelope built up from expanded van der Waals spheres centered on each of the atoms (Figure 7). A window containing a uniform grid of points is placed at a given distance from the surface of the molecule. For the points where the perpendiculars of the grid touch the surface envelope, the MEP of the molecule is calculated. In this way, we can obtain a two-dimensional representation of the MEP on the three-dimensional surface envelope of the molecule. For calculating the MEF which corresponds to the electrostatic interaction energy of a molecule with a point dipole, a similar representation is used.

The MEP of the most stable conformation of thiamin ylid is shown in Figure 8. The curves represent the isoenergy lines in kilojoules per mole. The areas with negative electrostatic potentials in the moiety of the thiazolium and pyrimidine ring confirm the preferred positions of the attack of a Mg^{2+} ion. For comparison we have also calculated the MEP of that conformation of thiamin ylid (Figure 9) which is generated by the influence of a Mg^{2+} ion above the thiazolium ring. The conformational changes of the molecule caused by the influence of the Mg^{2+} ion produce a remarkable increase of the negative potential above the thiazolium ring that is the attacking position of the cation. Analogous studies were performed on TPP and TPP ylid (TPP$^-$) in order to investigate the role of the pyrophosphate group.

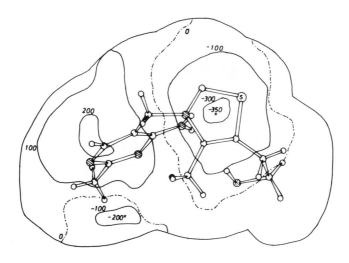

FIGURE 8.   MEP of thiamin ylid (1).

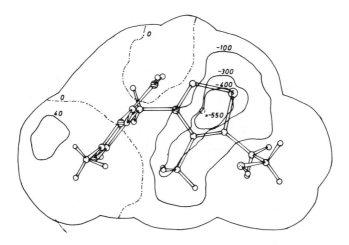

FIGURE 9.   MEP of thiamin ylid [Mg^{2+}] (TH).

Starting from the conformational maps, a complete variation including seven torsion angles gives the preferred conformations summarized in Table 2. Both the stable TPP and TPP ylid conformers have nearly the same ring orientations, which correspond with the S form. The F form found in the crystal structure of TPP is significantly less stable. In contrast to the β-hydroxyethyl group in thiamin, the longer side chain in TPP has a smaller flexibility. The most stable conformation of TPP is characterized by an extended arrangement of the side chain (Figure 10). It differs essentially from the orientation in the crystallographically found structure, where the side chain is back-folded above the thiazolium ring (Figure 11). In the conformer TPP (3) (Table 2) we have found a ring structure of the side chain forming a S1−H−O hydrogen bonding. A structure with a folded arrangement of the side chain results for the most preferred conformation of TPP ylid (Figure 12).

The MEP and the component of the MEF perpendicular to the surface envelope of TPP (1) (Figures 13 and 14) indicate essentially three preferred areas of the attack by an electrophilic agent: the pyrophosphate group, the pyrimidine, and the thiazolium ring. Therefore, we have considered as possible locations for the attacking of a Mg^{2+} ion to TPP and TPP ylid the positions above the centers of the two rings and above the center of the triangle

Table 2
TORSION ANGLES (IN DEGREES) AND RELATIVE ENERGIES (kJ/mol) OF
THE PREFERRED CONFORMATIONS OF TPP AND TPP YLID

Conformation	φ_T	φ_P	φ_1	φ_2	φ_3	φ_4	φ_5	E_r
TPP								
(1)	−129.4	−143.2	−4.8	43.9	−130.2	−116.5	82.2	0
(2)	−129.2	−144.3	60.9	−33.2	−75.5	−117.5	80.6	36
(3)	−129.2	−145.3	−87.2	117.4	−54.2	−109.6	49.2	95
Crystal structure	2.7	93.1	92.2	−62.9	140.9	−132.0	−132.7	338
TPP⁻								
(1)	−131.5	−145.3	−32.3	79.1	38.6	−116.3	−49.7	0
(2)	−127.8	−147.2	131.4	102.8	24.9	−109.9	−48.9	20
(3)	−128.5	−148.6	−30.9	77.9	39.3	−20.0	−48.9	22

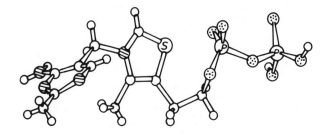

FIGURE 10.   TPP (1).

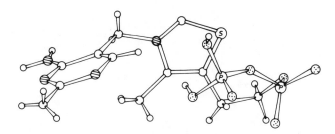

FIGURE 11.   TPP (crystal structure).

built up by the three oxygen atoms O10, O11, and O14 (Figure 1) of the pyrophosphate group. These three attacking positions are distinguished by the abbreviations TH, PY, and PP. In the most favored structure of the Mg^{2+}-TPP complex the cation is positioned 1.9 Å above the center of the pyrimidine ring and the relaxed interaction energy is −1895 kJ/mol (Figure 15). By the attack of the Mg^{2+} ion, the side chain of the TPP molecule is folded in such a way that an additional coordination of the cation by three oxygen atoms of the pyrophosphate group takes place. The Mg^{2+} ion-O distances (R_{Mg-O10} = 1.9 Å, R_{Mg-O11} = 2.0 Å, R_{Mg-O14} = 3.4 Å) can be compared with bond lengths crystallographically found in Mg^{2+} peptide complexes and in MgO minerals, as well.[18] Moreover, the Mg^{2+} ion causes a remarkable change with respect to the orientation of the rings. By the complexation of the Mg^{2+} ion to TPP a V-like structure is generated which is assumed to be important in the enzymatic mechanism (Figure 16).[19] In comparison with the most favored Mg^{2+}-TPP (PY) complex the Mg^{2+}-TPP (PP) structure is less stable by only 18 kJ/mol. In contrast to the other positions considered, the Mg^{2+} complexation of TPP above the center of the

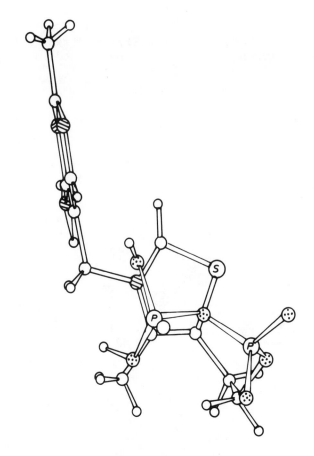

FIGURE 12.   TPP ylid (1).

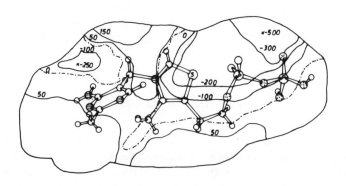

FIGURE 13.   MEP of TPP (1).

thiazolium ring is significantly less favored. The Mg^{2+} ion is positioned 4.3 Å above the thiazolium ring and the side chain is folded so that the pyrophosphate group can interact with the cation (Figure 17). The orientation of the two rings in the Mg^{2+}-TPP (TH) complex differs significantly from the V-like structure.

Analogous studies of the interaction of a Mg^{2+} ion to TPP ylid are performed. The results of the Mg^{2+} complexes of TPP ylid and TPP are compared in Table 3. The changed charge distribution in TPP ylid favors the attack of the cation above the center of the thiazolium ring, and the amounts of the interaction energies are generally larger in the Mg^{2+}-TPP ylid

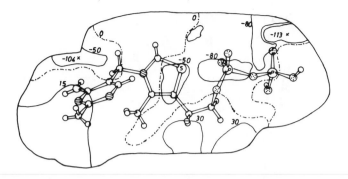

FIGURE 14.   MEF of TPP (1) (component perpendicular to the surface envelope).

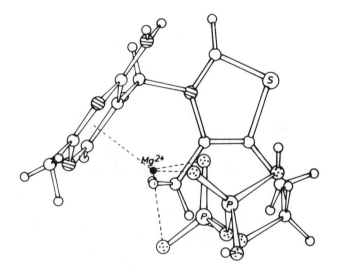

FIGURE 15. Mg^{2+}-TPP (PY).

complexes. The preferred structures are represented in Figures 18 to 20. In all cases we found V-like orientations of the two rings and the oxygen atoms of the pyrophosphate group take place on the coordination of the cation similar to that found in the Mg^{2+}-TPP complexes. The electrostatic potential of that conformation of TPP generated by the attack of a Mg^{2+} ion on the pyrimidine ring is shown in Figure 21. A comparison of Figures 13 and 21 illustrates the significant changes with respect to the conformation and electrostatic potential of the TPP molecule caused by the Mg^{2+} ion. The relative energies of the relaxed conformations of thiamin ylid, TPP, and TPP ylid found in the corresponding Mg^{2+} complexes are listed in Table 4. The values show that the interaction of Mg^{2+} ions generates conformations of the thiamin ylid, TPP, and TPP ylid molecules which are separated from the most stable ones by relatively high energy barriers. In the Mg^{2+} complexes these conformational barriers are overcompensated by the interaction energies. Although our results about the preferred complexation of a Mg^{2+} ion to thiamin ylid, TPP, and TPP ylid were obtained for the isolated molecules, in some ways they can be related to NMR studies on thiamin systems with paramagnetic cations in solution.[4-7] The shifting and broadening of the PMR and CMR signals caused by the cations indicate similar interactions involving the pyrophosphate group and the ring moieties. Thus, our investigations may give some further insights into the conformational behavior of the thiamin systems and its interaction with a Mg^{2+} ion from a theoretical point of view.

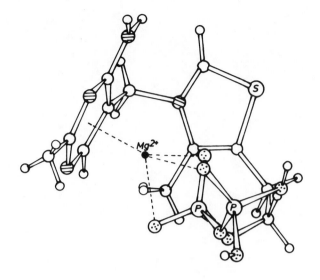

FIGURE 16. Mg^{2+}-TPP (PP).

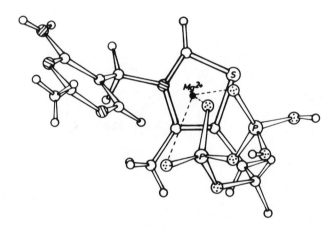

FIGURE 17. Mg^{2+}-TPP (TH).

Table 3
INTERACTION ENERGIES (kJ/mol) AND DISTANCES
(Å) OF Mg^{2+} COMPLEXES WITH TPP AND TPP YLID

Attacking position of cation	Mg^{2+}-TPP	Mg^{2+}-TPP ylid
PY[a]	−1896 (1.9)	−2430 (2.1)
TH[b]	−1721 (4.3)	−2523 (2.2)
PP[c]	−1878 (1.8, 2.2, 2.5)[d]	2355 (1.8, 2.2, 2.5)[d]

Note: Distances in parentheses.

[a]   Above the center of the pyrimidine ring.
[b]   Above the center of the thiazolium ring.
[c]   Moiety of pyrophosphate group.
[d] $R_{Mg^{2+}-O10}$, $R_{Mg^{2+}-O11}$, $R_{Mg^{2+}-O14}$; see Figure 1 and text.

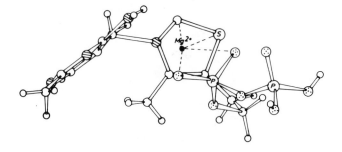

FIGURE 18. Mg^{2+}-TPP ylid (TH).

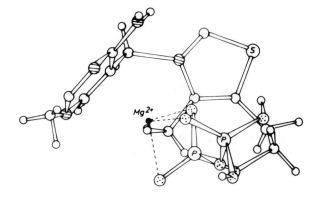

FIGURE 19. Mg^{2+}-TPP ylid (PY).

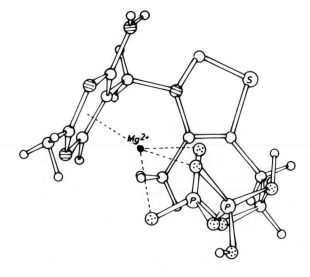

FIGURE 20. Mg^{2+}-TPP ylid (PP).

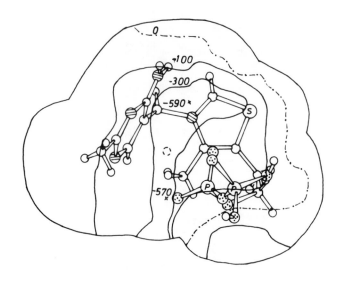

FIGURE 21.   MEP of TPP[Mg^{2+}] (PY).

Table 4
RELATIVE CONFORMATIONAL ENERGIES (kJ/mol) OF THE RELAXED LIGANDS THIAMIN YLID, TPP, AND TPP YLID FOUND IN THE CORRESPONDING Mg^{2+} COMPLEXES

Attacking position of cation[a]	Thiamin ylid[Mg^{2+}]	TPP[Mg^{2+}]	TPP ylid[Mg^{2+}]
PY	204	572	627
TH	214	295	545
PP	—	564	612

Note: Energies are related to those of the most stable conformations of the isolated ligands.

[a]   See Table 3.

REFERENCES

1. **Pletcher, J. and Sax, M.,** *J. Am. Chem. Soc.,* 94, 3998, 1972.
2. **Pletcher, J., Sax, M., Sengupta, S., Chu, J., and Yoo, C.S.,** *Acta Crystallogr.,* B28, 2928, 1972.
3. **Shin, W., Pletcher J., Blank, G., and Sax, M.,** *J. Am. Chem. Soc.,* 99, 3491, 1977.
4. **Gallo, A. A., Hansen, I. L., Sable, H. Z., and Swift, T. J.,** *J. Biol. Chem.,* 247, 5913, 1972.
5. **Gallo, A. A. and Sable, H. Z.,** *J. Biol. Chem.,* 250, 4986, 1975.
6. **Jordan, F. and Mariam, Y. H.,** *Microchem. J.,* 22, 182, 1977.
7. **Petzold, D. R. and Fischer, G.,** *Stud. Biophys.,* 54, 53, 1976.
8. **Jordan, F.,** *J. Am. Chem. Soc.,* 96, 3523, 1974.
9. **Jordan, F.,** *J. Am. Chem. Soc.,* 98, 808, 1976.
10. **Scheffers-Sap, M. M. E. and Buck, H. M.,** *J. Am. Chem. Soc.,* 101, 4807, 1975.
11. **Friedemann, R. and Gründler, W.,** *Z. Phys. Chem. (Leipzig),* 265, 561, 1984.
12. **Friedemann, R. and Gründler, W.,** *Z. Phys. Chem. (Leipzig),* 265, 775, 1984.
13. **Snir, J. R., Nemenoff, R. A., and Scheraga, H. A.,** *J. Phys. Chem.,* 82, 2497, 1978.
14. **Scrocco, E. and Tomasi, J.,** *Curr. Top. Chem.,* 42, 95, 1973.
15. **Pullman, A. and Pullman, B.,** *Q. Rev. Biophys.,* 14, 289, 1981.
16. **Friedemann, R. and Brandt, W.,** *Z. Phys. Chem. (Leipzig),* 266, 1201, 1985.
17. **Lavery, R. and Pullman, B.,** *Int. J. Quant. Chem.,* 20, 259, 1981.
18. **Karle, I. L.,** *Int. J. Peptide Protein Res.,* 23, 32, 1984.
19. **Schellenberger, A.,** *Ann. N.Y. Acad. Sci.,* 378, 51, 1982.

Chapter 3

^{31}P- AND ^{13}C-NMR STUDIES ON THIAMIN PYROPHOSPHATE

Sabine Flatau and A. Zschunke

TABLE OF CONTENTS

I. ³¹P-NMR STUDIES ON TPP

³¹P-nuclear magnetic resonance (NMR) investigations of TPP and its analogs are well known, e.g., from the works of Chauvet-Mongue et al.[1] and Petzold and Storek.[2] These authors inquired into the ³¹P-NMR chemical shifts of TPP and its dependence on several factors such as pH and metal ions. Referring to these investigations, the chemical shifts of TPP in aqueous solution were compared with those of a crude enzyme solution of PDC. Figure 1 gives the well-known dependence of both phosphate residues in the range 1 to 8, characterizing the transition to the twofold negative pyrophosphate residue. ³¹P-NMR measurements with the PDC solution were performed at pH 5 to 7. Figure 2 displays both the spectrum of TPP in aqueous solution and of PDC solution.

The chemical shifts found for the enzyme-bound TPP were compared with those from the aqueous TPP solution. According to a recommendation of Petzold, an empirical susceptibility correction was performed. Inorganic phosphate ³¹P-chemical shifts in both aqueous and enzyme solutions were compared and the difference was ascribed to magnetic susceptibility. Table 1 presents the data obtained in this way. The values compiled indicate that the environment of the β-phosphate remains fairly constant, whereas the α-shift suggests a more acidic environment in the enzyme. This could support the assumption that the TPP molecule is attached within the enzyme via the β-phosphate residue. The α-phosphate residue more exposed to the microenvironment of the active center displays shift changes up to 3 ppm. Also, a changed pK_a value could give rise to that phenomenon. The PDC preparation was performed according to a modified prescription of Ullrich[3] without use of phosphate. During all measurements, the temperature was kept constant at 283 K. The enzyme solution contained D_2O and ammonium sulfate. Enzymatic activity was the same prior to and after the NMR measurements. A Bruker NMR spectrometer WP 200 with 81.3 MHz for ³¹P was used.

II. ¹³C-NMR RELAXATION RATE MEASUREMENTS OF THIAMIN AND TPP

These investigations were performed previously and presented by Gallo and Sable[4] at the first TPP conference in 1981. They referred to a relation, widely used in NMR, namely:

$$1/NT_1 \propto \tau_c$$

where N is the number of directly attached protons; T_1 is the spin-lattice relaxation time; and τ_c is the correlation time. This value of NT_1 was used by Gallo and Sable to characterize the flexibility of the two heterocyclic rings and segmental motions within the whole molecule. We pursued this thread by applying an equation to the relaxation times of thiamin and TPP which describes the precise relation between the dipole-dipole relaxation rate of a magnetic nucleus and its correlation time:[5]

$$1/T_1^{DD} = n(\mu_o/4\pi)^2\gamma_H^2\gamma_c^2\hbar^2r_{C,H}^2\tau_c$$

where $1/T_1^{DD}$ = dipole-dipole relaxation rate; n = number of directly attached protons; μ_o = induction constant; $\gamma_{C,H}$ = magnetogyric ratio of C,H; $\hbar$ = $h/2\pi$; and $r_{C,H}$ = distance C–H. Utilizing this relation, we find the data presented in Table 2. As can be seen, the relaxation times of thiamin and TPP differ by one order of magnitude, reflecting the difference in molecular versatility to be of the same magnitude. The latter is described by the correlation times.

The C atoms chosen for this calculation represent the two heterocyclic rings. Within one molecule, the correlation times are of the same order of magnitude. This justifies the

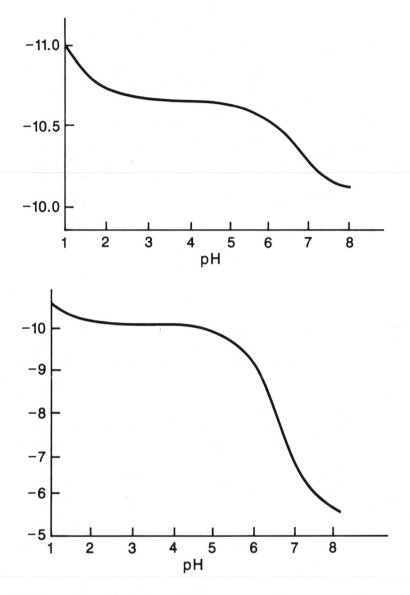

FIGURE 1.   pH dependence of chemical shifts. Upper graph: α-phosphate of TPP; lower graph: β-phosphate of TPP. TPP 0.2 *M*/D$_2$O, 298 K.

application of the modified Solomon-Bloembergen equation.[6-10] This equation describes the quantitative relations between the innersphere relaxation rate of a nucleus within a paramagnetic substrate-metal complex and the structural features of this complex.

$$T_{1M}^{-1} = \frac{2}{15} \frac{S(S + 1)}{r^6} \gamma_I^2 g^2 B^2 \left(\frac{3\tau_c}{1 + \omega_I^2 \tau_c^2} + \frac{7\tau_c}{1 + \omega_s^2 \tau_c^2} \right)$$

$$+ \frac{2}{3} \frac{S(S + 1) A^2}{\hbar^2} \left(\frac{\tau_e}{1 + \omega_s^2 \tau_e^2} \right)$$

where: S = electron spin quantum number; T_{1M} = innersphere relaxation time; γ_I = magnetogyric ratio of the nucleus I; r = distance metal ion nucleus; g = electronic g factor;

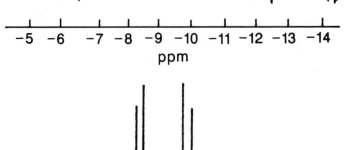

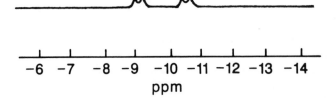

FIGURE 2.   Upper graph: ^{31}P-NMR spectrum of a PDC solution, pH 6.0, 5 × 10^{-4} *M*; lower graph: ^{31}P-NMR spectrum of a TPP solution, pH 6.0, 0.2 *M*/D$_2$.

Table 1

pH	$\Delta\beta_P$	$\delta\alpha_P$
5.0	−9.8	−10.4
6.0	−9.7	−12.2
7.0	−9.2	−13.5

Table 2

	Thiamin		TPP	
C atom	T_1(s)	(s)	T_1(s)	(s)
6′	1.2	3.9×10^{-11}	0.5	1×10^{-10}
2′	10.4	7.8×10^{-11}	2.45	3.3×10^{-10}
4	16.1	5.35×10^{-11}	3.3	2.6×10^{-10}
5	15.8	8.6×10^{-11}	4.5	3.0×10^{-10}

Table 3
RELAXATION DATA AND RELATIVE
DISTANCES OF THIAMIN (0.5 *M*/D$_2$O, pH 5,
MnSO$_4$ 5 × 10^{-4} *M*)

C atom	δ (ppm)	T$_{1F}$	T$_{10BSD}$	r$_i$
2'	163.9	10.41	1.52	1.44 ± 0.15
4'	163.7	12.67	1.78	1.48 ± 0.09
2	155.5	0.58	0.31	1.20 ± 0.10
6'	145.4	1.19	0.19	1.0 (reference)
4	143.5	16.07	2.15	1.53 ± 0.14
5	137.3	15.77	1.70	1.45 ± 0.07
5'	107.0	13.17	1.75	1.47 ± 0.07
5β CH$_2$	61.0	1.62	0.63	1.30 ± 0.07
5' CH$_2$	50.6	1.66	0.55	1.24 ± 0.07
5α CH$_2$	30.1	1.63	0.73	1.36 ± 0.07
2' CH$_3$	21.9	3.27	1.33	1.51 ± 0.08
4 CH$_3$	12.0	1.59	1.40	—

β = Bohr magneton; ω$_{I,S}$ = Larmor precession frequency for nucleus and electron spin; A = hyperfine coupling constant; τ$_c$ = correlation time of the dipolar nucleus-electron interaction; τ$_C$ = correlation time of the hyperfine interaction; and ħ = h/2. The first part of this equation involves the dipolar contribution; the second part the scalar contribution to the relaxation rate. The latter is negligibly small in the presence of paramagnetic ions, such as Mn^{2+}. Thus, and allowing for other preconditions,[6,8,10] this Solomon-Bloembergen equation can be modified. On understanding that all C atoms within the molecule possess nearly the same correlation time, ratios of T$_{1M}$ can be used to determine ratios of the distances between C atoms and metal ions, hence to determine relative distances:

$$T_{1Mi}^{-1}/T_{1Mj}^{-1} = r_j^6/r_i^6$$

Relaxation rates were measured in the presence and absence of Mn^{2+} ions. A so-called outersphere contribution has to be taken into account. This was calculated by determining the relaxation rate of the nucleus which shows the smallest influence of the paramagnetic ion in its relaxation rate. Table 3 presents the results.

The relative distances between Mn^{2+} and single C atoms were used to build calotte models. It thereby became evident that, for the thiamin-Mn^{2+} complex, no conformation is possible corresponding to all distances obtained. The distances therefore represent mean values of different thiamin-Mn^{2+} complexes in which the Mn^{2+} ion interacts first with the OH group and second with the two heterocyclic rings folded against each other by about 80°.

In the latter case, the shortest distances are found for C6', C2, and 5 CH$_2$, thus indicating that the metal ion should reside preferentially contiguous to the pyrimidine ring near the N1 atom. Because of the short lifetime of the thiamin-Mn^{2+} complex, the presence of two Mn^{2+} ions in the solvation sphere of the thiamin molecule is very unlikely. A calotte model formed on the basis of the calculated distances for the TPP-Mn^{2+} complex (Table 4) permits the following conclusions:

1. The TPP-Mn^{2+} possesses a preferred conformation in aqueous solution in which the pyrophosphate moiety folds upon the thiazolium ring.
2. The first coordination sphere of the complex contains, besides the PO groups of the pyrophosphate residue, the thiazolium ring of thiamin. The Mn^{2+} ion appears to be embedded between pyrophosphate and the thiazolium ring.

Table 4
RELAXATION DATA AND RELATIVE DISTANCES OF TPP (0.2 M/D$_2$O, pH 2, MnSO$_4$ 2 × 10^{-4} M)

C atom	δ (ppm)	T_{1F}	T_{1OBSD}	r_i
2'	163.8	2.45	0.41	1.60 ± 0.06
4'	163.6	4.62	0.46	1.63 ± 0.06
2	155.4	0.50	0.64	1.23 ± 0.07
6'	145.3	0.49	0.24	1.56 ± 0.08
4	143.4	3.30	0.22	1.28 ± 0.06
5	137.2	4.47	0.19	1.23 ± 0.06
5'	106.9	4.06	0.46	1.63 ± 0.07
5β CH$_2$	60.9	0.32	0.01	1.09 ± 0.07
5' CH$_2$	50.5	0.31	0.22	—
5α CH$_2$	30.0	0.34	0.13	1.23 ± 0.07
2' CH$_3$	21.8	1.28	0.39	1.74 ± 0.07
4 CH$_3$	11.9	1.37	0.07	1.00 (reference)

The C atoms of the thiazolium ring have distances shorter than those of the pyrimidine ring; therefore, the distances within one ring resemble each other.

TPP C atom	r_{rel}	
4 CH$_3$	1.0	
5β CH$_2$	1.23	
2	1.23	
5	1.23	Thiazolium ring
5α CH$_2$	1.23	
4	1.28	
6'	1.56	
2'	1.60	Pyrimidine ring
4'	1.62	
5'	1.63	

The influence of H bridges was not investigated.

Previous ^{1}H-NMR investigations of TPP by Grande et al.[11] revealed a structure somewhat different from that of our measurements, especially concerning the location of the metal ion. However, both investigations agree in requiring that the pyrophosphate residue fold upon the thiazolium ring, and in inferring that the structure explored in aqueous solution is very different from that within the active center.

III. EXPERIMENTAL CONDITIONS

^{13}C-NMR spectra were recorded on NMR spectrometer WP 200 of the Bruker Analytik Meβtechnik firm at 50.32 MHz with proton broadband decoupling at 303 K. T_1 values were determined by the inversion recovery method and automatically calculated from eight experiments. Thiamin and TPP were used as thiamin-HCl (purest) and TPP-HCl (purest) from the Serva firm. Prior to the T_1 experiments, samples were degassed (5× pump-freeze-thaw cycles).

REFERENCES

1. **Chauvet-Mongue, A. M., Hadika, M., Grevat, A., and Vincent, E. J.,** *Arch. Biochem. Biophys.,* 107, 311, 1980.
2. **Petzold, D. R. and Storek, W.,** *Stud. Biophys.,* 75, 1, 1979.
3. **Ullrich, J.,** *Meth. Enzymol.,* 18A, 109, 1980.
4. **Gallo, A. A. and Sable, H. Z.,** *Ann. N. Y. Acad. Sci.,* 378, 78, 1982.
5. **Lambert, J. B., Nienhuis, R. J., and Keepers, J. W.,** *Angew. Chem.,* 93, 553, 1981.
6. **Mildvan, A. S. and Gupta, R. J.,** *Meth. Enzymol.,* 49B, 322, 1978.
7. **Mildvan, A. S., and Engle, J. L.,** *Meth. Enzymol.,* 26C, 654, 1972.
8. **Lee, J. Y., Hanna, D. A., and Everett, G. W.,** *Inorg. Chem.,* 20, 2004, 1981.
9. **Swift, T. J. and Connick, R. E.,** *J. Chem. Phys.,* 37, 307, 1962.
10. **Lee, J. Y., and Everett, G. W.,** *J. Am. Chem. Soc.,* 103, 5221, 1981.
11. **Grande, H. J., Houghton, R. L., and Veeger, C.,** *Eur. J. Biochem.,* 37, 563, 1973.

Chapter 4

REACTIVITY AND MOLECULAR STRUCTURE OF THIAMIN PYROPHOSPHATE STUDIED BY NMR

Dieter R. Petzold

TABLE OF CONTENTS

I. INTRODUCTION

The reactivity of TPP is determined by its chemical structure which consists of three distinct moieties: the pyrimidine ring containing the 1'-nitrogen and the 4'-amino group, the thiazolium ring with the reactive 2–CH group, and the pyrophosphate residue (Figure 1).

Both heteroaromatic rings of TPP dissolved in water show a high mobility evidenced by ^{13}C relaxation measurements.[1] This means that the conformers of TPP and thiamin[1-4] exist in an equilibrium state. The pyrophosphate moiety and bivalent metal ions reduce the conformational mobility[1] and influence the conformational state.[5]

The 2–C–H bond is characterized by its marked lability which is reflected in a relatively short lifetime of the 2–H in aqueous solution.[6-8] The intermediate carbanion of TPP bound to the enzyme is able to react with substrates. The equilibrium between the protonated and carbanion form of TPP dissolved in water is controlled by the hydroxyl ion concentration. A similar equilibrium was found for thiamin and oxythiamin,[7-9] but these compounds are without coenzymatic activity.

In this chapter, molecular properties of TPP and its derivatives as studied by NMR are compared with their reactivity monitored by the lifetime of the 2–H. Since the reprotonation of the carbanion is a fast ionic reaction,[10] the lifetime of the 2–H decides the equilibrium state of both molecular forms.

II. MATERIALS AND METHODS

All compounds used were of analytical grade. The pH values measured by a combined glass electrode (Radiometer, Denmark) were adjusted using 20% hydrochloric acid for pH < 3, sodium bicarbonate for 3 < pH < 7, and sodium hydroxide following the bicarbonate treatment with intensive stirring for pH > 7. The stability of the dissolved compounds was checked by their NMR spectra. Contrary to Reference 11, normal water was used as solvent in all experiments to observe the signals of both 2–H and 2–C. NMR spectra were run at a magnetic field strength of 2.347 T on a PFT 100 spectrometer (Jeol, Japan) in the FT mode at 23°C for ^{13}C and ^{31}P and on a KRH 100 spectrometer (Workshop of the Academy of Sciences of the German Democratic Republic) in CW mode at 27°C for ^{1}H. EPR spectra were run at 25°C and pH 6.5 on an E3 spectrometer (Varian, U.S.).

III. RESULTS AND DISCUSSION

The dependence of carbon, phosphorus, and proton chemical shifts of TPP on the pH is demonstrated in Figure 2. The phosphorus spectra indicate that pK = 6.3 for the pyrophosphate moiety which is not reflected in the carbon or proton spectra. These spectra demonstrate 1'–N as the protonation site with pK = 5.3 and 5.0 for TPP and thiamin, respectively. Only the signals of the aliphatic chain and 5–C were shifted markedly by the pyrophosphate.

The reliability of using chemical shifts as monitors for electronic effects is proved by correlations of them with electron net charges of the corresponding carbon atoms calculated for thiamin by CNDO/2 method.[13] In Figure 3, good linear least-squares fits for two pH values are demonstrated for the chemical shifts of the aliphatic carbons of TPP with their net charges (correlation coefficients r > 0.98). The slope for the pyrimidine carbons has the same sign and magnitude as for the aliphatic carbons (theoretical value approximately equals 160 ppm/e). This means that the differences of the net charges of the pyrimidine ring carbons represent mainly different σ-electron densities.

It is noticed that for TPP and thiamin, the proton signals of the 4'-amino group and the

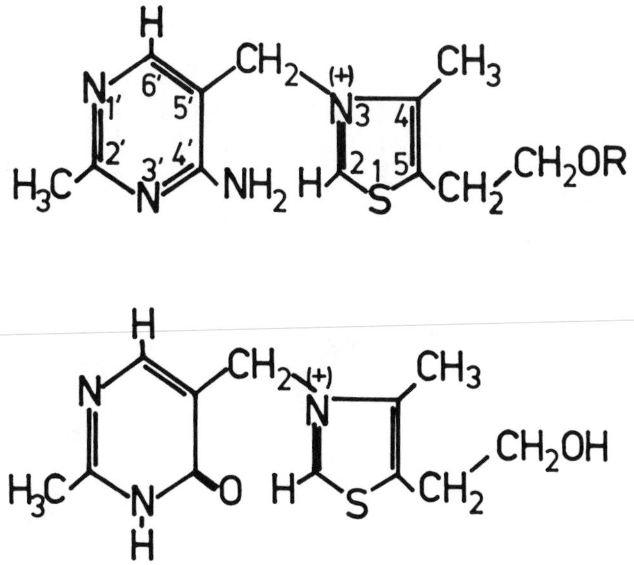

FIGURE 1. Chemical structures of TPP and derivatives used. Above: R = $P_2O_6^{3-}$: TPP; R = H: thiamin. Below: oxythiamin.

2–H but not the carbon signals of the 4′–C and of the 2–C are influenced by the 1′–N protonation (Figure 2). The 1′-proton is attached to the 1′–N lone pair and perturbs in this way the distribution of the π-electron density of the pyrimidine ring. Protonation-dependent decrease of the π-electron density produces high field shifts of the carbon signals of 6′–C, 2′–C, 2′–CH₃ and the corresponding low field shifts of the proton signals of 6′–H, 2′–CH₃, and the amino group. The influence of the 1′–N lone pair on the π-electron density of the 6′–C and 2′–C is confirmed by the separate straight line of both carbons in Figure 3a. Their higher δ values indicate a higher π-electron density compared to the 1′–N protonated molecule. A similar but smaller effect is shown for 2′–CH₃ (Figure 3b, c). The signal of the 2–H is shifted by anisotropic effects originated in a changed π-electron density in the pyrimidine ring, especially in the neighboring amino group. Direct electronic effects via methylene bridge, 3–N, and 2–C do not contribute to the 2–H chemical shift changes because the signals of the 2–C and the methylene bridge carbon do not show comparable shifts. pH-dependent differences of the conformational state connected with alterations of the anisotropic influence of the pyrimidine ring on the 2–H can be excluded for thiamin because its vicinal carbon proton coupling constants $^3J_{2-C/3-H_2}$ and $^3J_{6'-C/3-H_2}$ remain constant (Table 1). On the other hand, TPP undergoes a pH-dependent change of its conformational equilibrium state, suggesting interactions of the positively charged protonated 1′–N with the negatively charged pyrophosphate moiety that is immobilizing the conformational flexibility. However, the chemical shifts depend on the pH in the same manner for TPP as for thiamin (Figure 2).

This means that the pH-induced chemical shift changes originate for the same reason. Therefore, the restricted mobility caused by the pyrophosphate does not perturb the conformer which determines the proximity of the 2–H to the amino group.

The lifetime of the 2–H as monitor for the reactivity of TPP can be determined by proton NMR. Using D_2O as solvent, kinetics of the proton-deuterium exchange were measured directly in the range pD < 6 from the decreasing intensity of the 2–H proton signal depending on time.[6,8,9] Using normal water as solvent, the lifetime τ of the 2–H and its half-life time $t_{1/2}$ can be calculated from line broadening ΔLw of its NMR signal[14] in the range pH > 6.7:[7]

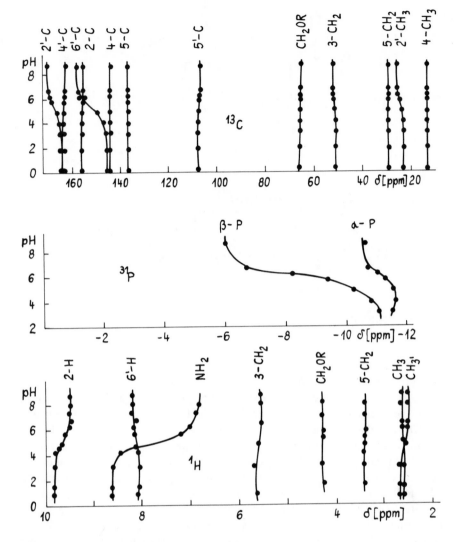

FIGURE 2. ^{1}H-, ^{31}P-, and ^{13}C-NMR chemical shift values (δ) for TPP in dependence on pH. Assignments were described in Reference 12.

$$\ln 2 \cdot (t_{1/2})^{-1} = \tau^{-1} = \pi\Delta Lw = \pi(Lw - Lw^\circ) \tag{1}$$

where Lw = measured line width of the 2–H, and Lw$^\circ$ = line width without exchange, e.g., of the 6′–H.

Logarithmic plots of the lifetimes determined for TPP, thiamin, and oxythiamin vs. pH are illustrated in Figure 4. The slopes of the linear least-squares fits for alkaline solutions amount to 0.6 for TPP and thiamin and 0.75 for oxythiamin, and characterize the reaction order[8] corresponding to the following equations:

$$0.4343 \ln \tau - 0.1592 = \log t_{1/2} = -0.6 \text{ pH} + k' \tag{2}$$

$$\frac{dc}{dt} = k \cdot [OH^-]^{0.6} \tag{3}$$

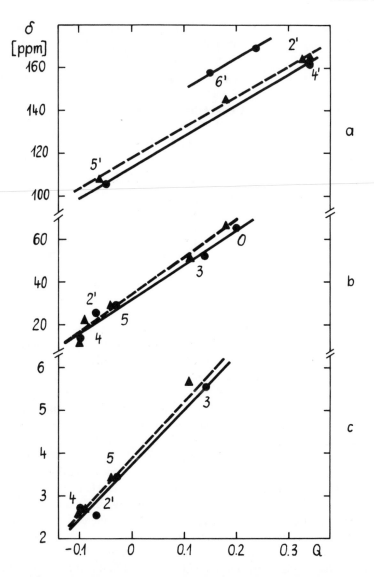

FIGURE 3.   Correlations of chemical shift values (δ) of TPP with electron net charges (Q) of the corresponding atoms[13] by least-squares fits: (●) pH 6.7; (▲) pH 1.9. (a) Carbon atoms of the pyrimidine ring; (b) aliphatic carbons (O: CH_2OR); (c) δ: aliphatic protons and Q: aliphatic carbon atoms to which the protons are bound.

The fractional exponents less than 1 indicate a catalytic mechanism for the dissociation of the 2–C–H bond. The catalytic mechanism is more effective for TPP and thiamin than for oxythiamin. Since the lifetimes measured for thiamin and TPP are independent of their concentration, the catalytic mechanism must be intramolecular.[15] The decreasing exchange rate measured for increasing concentrations of oxythiamin reveals that self association of oxythiamin perturbs the H–H exchange.

Comparing the 2–H life times in acidic and alkaline solution, it was found that the extrapolated straight lines for oxythiamin are parallel (Figure 4). Their short distance (Δlnτ = 1.1) is attributed to isotopic effects (H–H/H–D, pH/pD) and lies in the same range as found for H–T exchange.[10] The diffusion-controlled reprotonation increases the probability that the lost H bonds to the 2–C again instead of D of the solvent D_2O. In contrast to

Table 1
SELECTED VICINAL
$^{13}C-^1H$ COUPLING
CONSTANTS [Hz][12]
AS INDICATORS FOR
CONFORMATIONAL
CHANGE OF TPP
AND THIAMIN

	TPP		Thiamin	
	pH 1.9	pH 6.7	pH 2.3	pH 6.2
$^3J_{6'-C,3-H_2}$	4.4	6	5	5
$^3J_{2-C,\ 3-H_2}$	3.3	10	3	3

oxythiamin, the straight lines for TPP and thiamin are not parallel for acidic and alkaline solution. An attempt to connect the least-squares fits is shown by a nonlinear interpolation (dotted lines). Within experimental errors of such an interpolation, the inflection points of the dotted lines coincide with pK values of 1'−N protonation. Because its pK = 2.3,[9] an inflection point for oxythiamin can only appear outside of the measured pH range.

A proposal for the catalytic mechanism is shown in Figure 5. As found using space-filling molecular models, it was demonstrated that reaction steps are conceivable: an indirect proton transfer via a water molecule (Figure 5a) and the direct transfer (Figure 5c). Both catalytic mechanisms result in a further polarization of the 2−C−H bond induced by the electronegativity of the amino group, and, in the same way, a local increasing of the hydroxyl ion concentration which explains the calculated reaction order less than one. Decreasing electron density of the amino group influenced by 1'−N protonation and monitored by the low field shift of the amino proton signal is connected with a drastic increasing of the lifetime of the 2−H up to 15 times. Therefore, the 4'-amino group accelerates the dissociation of the 2−C−H bond and is in this way essential to the coenzymatic activity of TPP.

The minor exchange rate measured for TPP compared to thiamin can be explained by its molecular structure. The pyrophosphate moiety is folded in aqueous solution in such a way across the molecule that its negative charge is located near the pyrimidine ring.[16] By this pyrophosphate-pyrimidine interaction, the electron density distribution in the pyrimidine ring can be altered or the formation of the conformer catalyzing the dissociation of the 2−H can be perturbed, resulting in a loss of catalytic activity of the amino group.

The TPP is bound to the apoprotein of the pyruvate decarboxylase (PDC) via a Mg^{2+} ion attached to the 1'−N.[2,17] The holoenzyme PDC can be activated by pyruvate and pyruvamide.[18,19] Therefore, the influence of Mg^{2+} ions, pyruvate, and pyruvamide on lifetime and chemical shift of the 2−H was studied.

Complexes between Mg^{2+} ions, pyruvate, and thiamin exist already without the apoprotein. The complex between thiamin and pyruvate was proved using Mn^{2+} ions as bivalent cations. Free Mn^{2+} ions are easy to quantify by EPR in the form of their hexahydrates. Signals of asymmetrically liganded Mn^{2+} ions are quenched by strong line broadening. Table 2 contains the concentration of the free Mn^{2+} ions and the equilibrium constants for the systems studied. Thiamin and pyruvate reduce the concentration of the free Mn^{2+} ions by 10 and 69 μM, respectively, forming binary complexes. For the ternary system, the con-

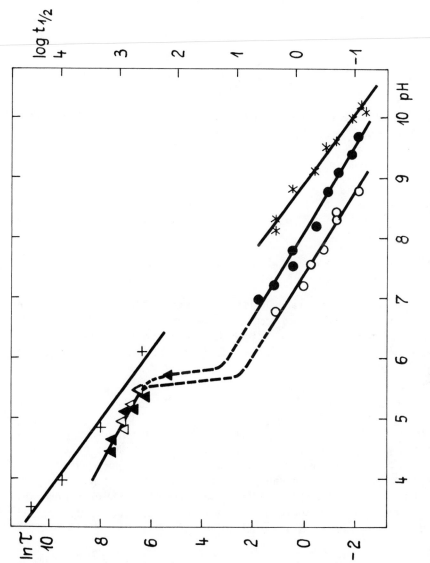

FIGURE 4.   Logarithmic plot of the lifetimes τ (s) of the 2–H vs. pH. TPP (▲) 28°C, H–D;[6] 28°C, H–D;[6] (●), 27°C, H–H.[15] Thiamin: (△) 28°C, H–D;[6] (○) 27°C, H–H;[15] Oxythiamin: (+) 30°C, H–D;[9] (*) 27°C, H–H.[15]

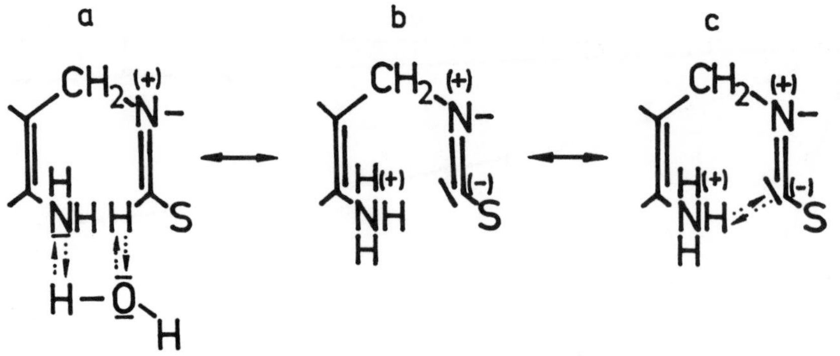

a   b   c

FIGURE 5.   Model of the catalytic effect of the 4'-amino group on the 2–H exchange.

Table 2
PARAMETERS OF THE COMPLEXES FORMED FROM THE COMPONENTS
THIAMIN (Th$^+$), Mn^{2+} IONS, AND PYRUVATE (Py$^-$)

System	c_{Mn}^{2+} (μM)	K (M^{-1})
Mn^{2+}	200	—
Th$^+$/Mn^{2+}	190	0.3
Py$^-$/Mn^{2+}	131	2.6
Th$^+$/Mn^{2+}/Py$^-$	153	1.5
Th$^+$/Py$^-$	—	10

Note: Thiamin and pyruvate: each 200 mM; $c_{Mn^{2+}}$: concentration of free manganese hexahydrate ions; K: constant of complex formation.

centration of the free Mn^{2+} ions was calculated to 124 μM provided that there is no interaction between thiamin and pyruvate. The measured higher concentration reveals a binary complex between thiamin and pyruvate (Th$^+$-Py$^-$). Its concentration depends on the amount of a possible ternary complex (Th$^+$-Mn^{2+}-Py$^-$) corresponding to the following equation[20] derived from the measured data:

$$c_{Th+Py-} = (0.10 \pm 0.04) + (2.2 \pm 0.3) \cdot 10^3 \, c_{Th+Mn^{2+}+Py-} \qquad (4)$$

From kinetics of the decomposition of pyruvate monitored by the decreasing intensity of the proton NMR signal of the pyruvate methyl group (Figure 6), a ternary complex between Mg^{2+} ions, pyruvate, and thiamin was evidenced. Pyruvate alone does not decompose significantly for the period of experiments. Mg^{2+} ions catalyze this decomposition better than thiamin. For the ternary system, only a little faster decomposition of the pyruvate than measured in the presence of Mg^{2+} ions is expected without formation of a ternary complex. The much faster decomposition indicates a ternary complex consisting of thiamin, Mg^{2+} ions, and pyruvate.

The effect of the interaction of Mg^{2+} ions, pyruvate, and pyruvamide with TPP and thiamin on the lifetime of the 2–H and its chemical shift is demonstrated in Figure 7. In complexes with the three effectors, the proton exchange is faster between 2.7 and 1.5 times compared to unliganded TPP and thiamin, respectively. A similar shortening of the lifetime of the 2–H (1.7 times) was found for a thiamin-indole complex.[22] It is assumed from the linearity of the demonstrated least-squares fits that Mg^{2+} ions, pyruvate, and pyruvamide influence the 2–H lifetime via electronic effects resulting from a stabilization of the catalytic conformation within the conformational equilibrium and electron density changes of the 2–CH and the amino group. The different sign of both slopes reveals a different mechanism of the influence on the electron density by the three effectors for TPP and thiamin.

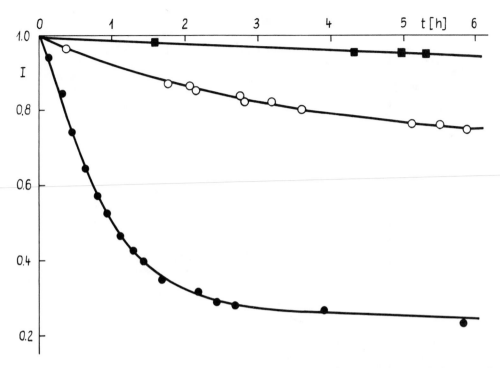

FIGURE 6.   Kinetics of the decomposition of pyruvate in dependence on Mg^{2+} ions and pyruvate. Each compound = 0.25 M in water at pH 6.5.[21] I: Intensity of the pyruvate methyl ^1H signal; t: time. (■) Pyruvate + thiamin; (○) pyruvate + Mg^{2+} ions; (●) pyruvate + thiamin + Mg^{2+} ions.

IV. CONCLUSIONS

Since ligands bound to TPP accelerate the dissociation of the 2–C–H bond, it is probable that TPP, which is firmly bound in the protein pocket and is liganded within by amino acid side chains as well as immobilized in respect to its conformational state, exists in a high degree as carbanion. A similar assumption was derived from kinetic comparisons.[10]

Regulation of the dissociation of the 2–C–H bond by electronic distribution effects on the 1′–N via the pyrimidine ring and the 4′-amino group has an analagon in the enzyme PDC. The 4′-amino group is involved in the enzymatic mechanism.[23] Effector binding to the regulatory center exerts conformational alterations of the protein.[19] Therefore, it is probable that the regulatory center of the PDC operates via conformational changes of the protein to the Mg^{2+} ion resulting in electronic changes at the 1′–N and via the pyrimidine ring at the amino group.

The results concerning the influence of the pyrophosphate moiety on the reactivity of the 2–C lead to the conclusion that the pyrophosphate is essential to the binding of TPP to the apoprotein only.

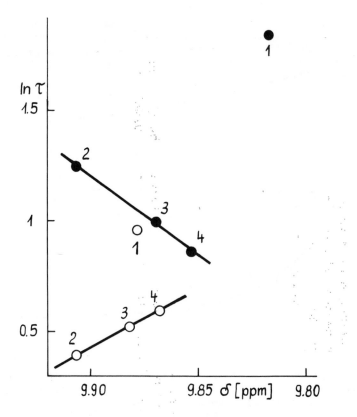

FIGURE 7.   Semilogarithmic plot of the lifetime τ vs. the ¹H chemical shifts δ of the 2–H of (●) TPP and (○) thiamin at pH 7. (1) Without ligand, (2) with Mg²⁺ ions, (3) with pyruvate, (4) with pyruvamide.

REFERENCES

1. **Gallo, A. A. and Sable, H. Z.,** *Ann. N.Y. Acad. Sci.,* p. 78, 1982.
2. **Schellenberger, A.,** *Angew. Chem. Int. Ed.,* 6, 1024, 1967.
3. **Shin, W., Pletcher, J., Blank, G., and Sax, M.,** *J. Am. Chem. Soc.,* 99, 3491, 1977.
4. **Cramer, R. E., Maynard, R. B., and Ibers, J. A.,** *J. Am. Chem. Soc.,* 103, 76, 1981.
5. **Chauvet-Monges, A.-M., Monti, J.-P., Crevat, A., and Vincent, E. G.,** *Biochimie,* 61, 1301, 1979.
6. **Chauvet-Monges, A.-M., Rogeret, C., Briand, C., and Crevat, A.,** *Biochim. Biophys. Acta,* 304, 748, 1973.
7. **Petzold, D. R.,** *Stud. Biophys.,* 54, 159, 1976.
8. **Ullrich, J. and Mannschreck, A.,** *Biochim. Biophys. Acta,* 115, 46, 1966.
9. **Suchy, J., Mieyal, J. J., Bantle, G., and Sable, H. Z.,** *J. Biol. Chem.,* 247, 5905, 1972.
10. **Kemp, D. S. and O'Brien, J. T.,** *J. Am. Chem. Soc.,* 92, 2554, 1970.
11. **Gallo, A. A. and Sable, H. Z.,** *J. Biol. Chem.,* 249, 1382, 1974.
12. **Petzold, D. R. and Storek, W.,** *Stud. Biophys.,* 75, 1, 1979.
13. **Jordan, F.,** *J. Am. Chem. Soc.,* 96, 3623, 1974.
14. **Forsen, S., Frankle, W. E., Laszlo, P., and Lubochinsky, J.,** *J. Magn. Reson.,* 1, 327, 1969.
15. **Petzold, D. R., Hübner, G., Fischer, G., Neef, H., and Schellenberger, A.,** *Stud. Biophys.,* 87, 15, 1982.
16. **Grande, H. J., Houghton, R. L., and Veeger, C.,** *Eur. J. Biochem.,* 37, 563, 1973.
17. **Ullrich, J. and Wollmer, A.,** *Hoppe-Seyler's Z. Physiol. Chem.,* 352, 1635, 1971.
18. **Hübner, G., Fischer, G., and Schellenberger, A.,** *Z. Chem.,* 10, 436, 1970.

19. **Nafe, G., Hübner, G., Fischer, G., Neef, H., and Schellenberger, A.,** *Acta Biol. Med. Ger.,* 29, 581, 1972.
20. **Petzold, D. R.,** *Stud. Biophys.,* 55, 231, 1976.
21. **Petzold, D. R. and Fischer, G.,** *Stud. Biophys.,* 54, 53, 1976.
22. **Ishida, T., Matsui, M., Inove, M., Hirano, H., Yameshita, M., and Sugiyama, K.,** *Biochem. Biophys. Res. Commun.,* 116, 486, 1983.
23. **Hübner, G., Neef, H., Fischer, G., and Schellenberger, A.,** *Z. Chem.,* 15, 221, 1975.

Chapter 5

GENERAL MECHANISM OF THE STRUCTURAL TRANSFORMATIONS OF THIAMIN IN AQUEOUS MEDIA

Jacques-Emile Dubois and Jean-Michel El Hage Chahine

TABLE OF CONTENTS

I. INTRODUCTION

Until quite recently, the intimate mechanisms governing the structural transformations of thiamin were not well known.[1,2] These transformations are not merely an interesting chemical problem, since this biological molecule is vital to the functioning of normal metabolic processes in life,[3] such as pyruvate decarboxylation and benzoin condensation, which occur before the start of the tricarboxylic acid cycle.[3]

On the basis of his research on the exchange of thiamin proton 2 with deuterons in heavy water, Breslow[4] was the first to postulate the existence of zwitterion A_{-H} as the biocatalytically active species in pyruvate decarboxylation. In our work, zwitterion A_{-H} is considered a resonance-stabilized carbene species[5] (Chart I), the action of which is illustrated by Schellenberger's[6] general scheme of pyruvate decarboxylation involving the joint action of thiamin and an enzyme (α-lactyl TPP) (Scheme I).

Breslow's catalyst is obtained by the deprotonation of thiamin proton 2. The negative charge on A_{-H} facilitates its addition to the carbonyl group because of the partial positive charge of this group. The resulting adduct protonates normally and deprotonates enzymatically to give the unaltered thiamin coenzyme and acetaldehyde (Scheme I).[6]

It would be overly simplistic to base any explanation of the biocatalytic activity of thiamin solely on this schematic system, because it describes only one precarious thiamin species and precludes all other structural modifications undergone by thiamin in aqueous media. According to Hopmann et al.,[7,8] thiamin deprotonates with a basic pK of 12.6. However, above pH 7, thiamin starts acquiring a very complex reactional behavior which would involve no less than 5 to 6 structural transformations, all of which are more privileged thermodynamically than deprotonation at position 2.

The transformation of thiamin into thiolate C^- occurs with two equivalents of OH^-; has an average pK of 9.3 (measured by Zoltewicz, Uray, Kluger, Chin, Duclos, Haake, Jordan, and many others);[9] and displays first-order kinetics. Nucleophile attachment can occur at thiamin position 2, so OH^- attachment would yield the thermodynamically underprivileged pseudobase B which, by thiazole ring opening and proton loss, would generate the thermodynamic product, thiolate C^-.[9]

In very basic media (pH > 11), the reactional behavior of thiamin becomes even more complicated, involving at least two thiamin structures other than A^+: B and C^-. Actually, a neutral colorless solution of thiamin, when rendered strongly basic, instantly turns yellow. This coloring disappears exponentially over time and yields a colorless solution containing thiolate C^-.

According to Maier and Metzler, thiamin in basic media rapidly generates the yellow form J^- via the σ-adduct D through ring opening with a proton loss (Chart I) resulting from the intramolecular nucleophilic addition of the amino group of aminopyridine at thiazolium ring position 2.[10] They measured an average pK of 11.6 for the $A^+ \rightleftarrows J^-$ transformation and also maintained that the $J^- \rightleftarrows C^-$ transformation evolves via σ-adduct D, cation A^+, and pseudobase B.[10]

The mechanism devised by Maier and Metzler was partly confirmed by Hopmann et al. who, using stopped-flow kinetics, showed that structural transformation of A^+ into J^- is rate limited by the $A^+ \rightleftarrows D$ step for which the rate constants have been measured.[8] However, they expanded this step by involving the deprotonation of A^+ into A_{-H}, believed responsible for the formation of σ-adduct D, which presumably engenders J^-.[7,8] To justify the involvement of A_{-H} in the $A^+ \rightleftarrows J^-$ transformation, they had to ascribe a pK of 14.15 to the deprotonation of cation A^+ into A_{-H}, in contradiction with the pK of 12.6 which they had always reported previously.[7,8] Later, on the basis of thermodynamic considerations, the same authors rejected Maier and Metzler's proposal concerning the indirect transformation of J^- into C^- via D, A^+, and B and instead proposed a direct transformation of J^- into C^-.[7]

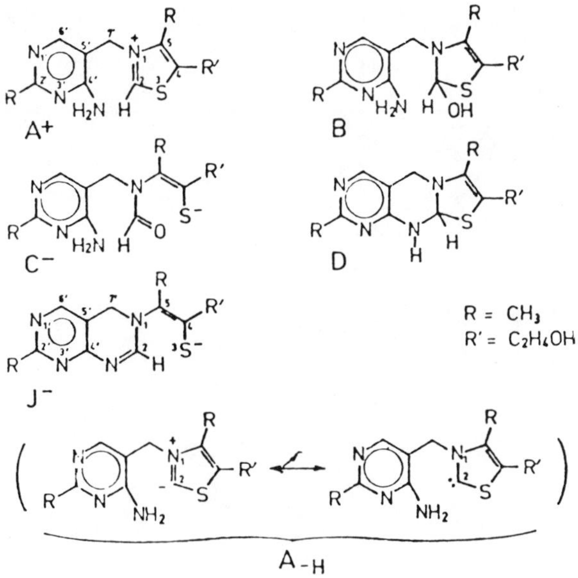

CHART I.

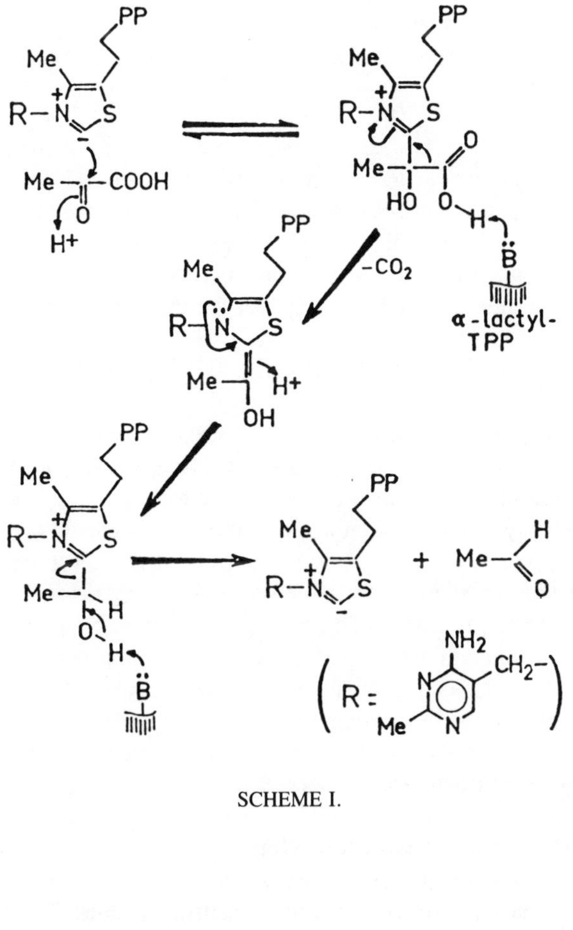

SCHEME I.

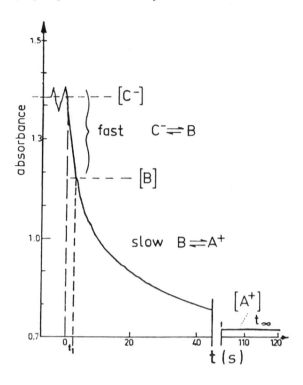

FIGURE 1.   Fast pH jump from basic (approximate pH 11.6) to
neutral (pH 6.9) at 25°C. Absorbance change of a thiamin solution
at 249 nm (thiamin concentration c $= 1.27 \times 10^{-4}$ M).

The work described herein focuses on the kinetics and thermodynamics of the structural
transformations of thiamin in aqueous media. The lability of thiamin proton 2 is also discussed. Chemical relaxation techniques and methods, especially pH jumps, which have
proved very helpful in understanding the complexity of the thiamin system, have been used
for kinetic data and to measure the UV spectra of the intermediate species.

II. STRUCTURAL TRANSFORMATIONS OF THIAMIN IN NEUTRAL MEDIA

If a basic solution of thiamin (containing thermodynamic product C^-) is rapidly neutra-
lized, two kinetic processes are observed at 249 nm: (1) a sharp drop in absorbance, followed
by (2) a slow exponential drop in absorbance (Figure 1).

Thiolate C^-, the only form of thiamin found in basic media, evolves in neutral media
after the pH jump into cation A^+, the only major form of thiamin found near neutrality. It
evolves in two successive steps: a fast first step which, in time t_1, corresponds to the
accumulation of a kinetic product which, in a second step, evolves with a relaxation time
τ_2 into cation A^+. A temperature drop slows the fast kinetic process which, over time,
becomes more noticeable.[1] This fast phenomenon is an exponential drop in absorbance with
a pH-independent amplitude in the pH range 6 to 8 and corresponds to the complete formation
and accumulation of pseudobase B, an intermediate between A^+ and C^-. We associate
relaxation time τ_1 to this fast relaxation mode.[1]

A. $C^- \rightleftarrows$ B Transformation (Relaxation Mode τ_1)

Near neutrality, the formation of pseudobase B from thiolate C^- can occur via two different
pathways: an acid-promoted and a base-promoted pathway (Figure 2). Experimental data fit
only the reciprocal relaxation associated with the base-promoted pathway (Figure 2).[1] The

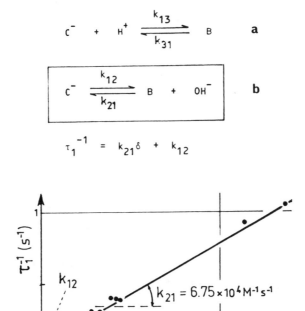

FIGURE 2. The structural transformations of thiolate in neutral media [(a) is the acid- and (b) is the base-promoted transformation]. Plot of $\tau_1^{-1} = k_{21} + k_{12}\,\delta$, where δ is a complex function of the concentrations of the species present in the medium.[1]

best linear least-squares regression gives rate constant k_{21} ($6.75 \times 10^4\ M^{-1}\ \text{sec}^{-1}$) and reverse rate constant k_{12}. The k_{12}/k_{21} ratio is equal to the equilibrium constant of the base-promoted transformation of pseudobase B into thiolate C^-.[1]

Hence, thiolate C^- is formed by the deprotonation and ring opening of pseudobase B.

B. B $\rightleftarrows$ A⁺ Transformation (Relaxation Mode τ_2)

From t_1 onward (Figure 1), the pseudobase accumulated by thiazole ring formation evolves into cation A⁺ with relaxation time τ_2.[1] Near neutrality, pseudobase B can evolve into cation A⁺ via two possible pathways: acid-promoted or base-promoted (Figure 3).[1] Thiamin is also known to protonate on the N−1′ atom of the pyrimidine with a pK of 5.22.[1] If this protonation is taken into account in the general expression of the reciprocal relaxation equation associated with the two possible pathways of A⁺ formation, experimental data will fit only the reciprocal relaxation equation associated with the acid-promoted dehydration of pseudobase B into cation A⁺ (Figure 3).[1] Therefore, it is possible to measure k_{23} ($1.15 \times 10^6\ M^{-1}\ \text{sec}^{-1}$) and k_{32}. The ratio of both rate constants is equal to k_H ($1.95 \times 10^{-10}\ M$), the equilibrium constant of pseudobase B formation from cation A⁺.

Near neutrality, acid-promoted dehydration transforms pseudobase B into cation A⁺.

III. STRUCTURAL TRANSFORMATIONS OF THIAMIN IN MILDLY BASIC MEDIA

What is the behavior of thiamin itself when placed in a medium with pH 9 to 10.5?

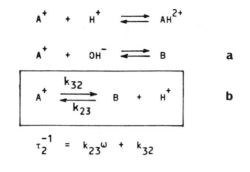

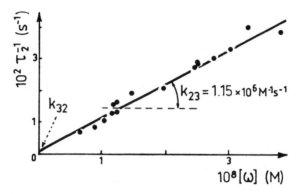

FIGURE 3.   The structural transformations of thiamin in neutral media [(a) is the acid- and (b) is the base-promoted transformation]. Plot of $\tau_2^{-1} = k_{23}\,\omega + k_{32}$, where ω is a function of the concentrations of the species present in the medium.[1]

When a neutral thiamin solution is rendered slightly basic, a single mode of relaxation reflected by an exponential drop in absorbance is detected at 233 nm.[1] Cation A^+, the only major thiamin species found in neutral media, is transformed into thiolate C^- by a unique kinetic step which does not involve the accumulation of any kinetic intermediate, but which is rate limiting.[1]

For the transformation of thiamin in basic media into pseudobase B by hydroxylation of thiamin site 2 (Figure 4),[1] B — as soon as it forms from thiamin — will change rapidly into thiolate C^- via the base-promoted pathway already seen in neutral media. Experimental data fit the reciprocal relaxation corresponding to the rate-limiting hydroxylation of cation A^+ (Figure 4).[1] The best linear least-squares regression gives 19.6 M^{-1} sec^{-1} for k_{34}, the hydroxylation rate constant.

In mildly basic media (pH < 11), thiamin is transformed into pseudobase B by the hydroxylation of thiamin position 2, the rate-limiting step in the structural transformations of thiamin.

IV. PRESENCE OF PSEUDOBASE B IN AQUEOUS MEDIA

From these results, the difference between the pK of the pseudobase cation equilibrium (pK$_H$ 9.70) and the pK of the thiolate/pseudobase equilibrium (pK$_T$ 8.90) is less than 1. This minor difference runs counter to a total absence of pseudobase from the reaction media. As a matter of fact, pseudobase B indeed exists to the extent of 16% of the overall thiamin concentration between pH 9.2 and 9.4.[1] Moreover, the UV spectrum of B has been measured indirectly at time t_1 (Figure 1), when B is the only species present in solution.[1]

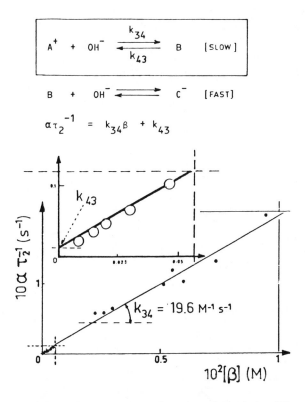

FIGURE 4. The structural transformations of thiamin in mildly basic (pH < 11) media. Plot of $\alpha \tau_2^{-1} = k_{43} + k_{34} \beta$, where α and β are complex functions of the concentrations of the species present in the solution.[1]

V. STRUCTURAL TRANSFORMATIONS OF THIAMIN IN BASIC MEDIA (pH < 11)

Were Breslow's proposal[4] about the deprotonation of A^+ into A_{-H} with a pK of 12.6 or 14.15 (as reported by Hopmann et al.)[7,8] to be taken into account, this would implicate that the biocatalytic activity of thiamin residues in its structural transformations in basic media. However, it should be noted that the pH of natural media is near neutrality and that the concentration of A_{-H} (if the pK of A^+ deprotonation is 12.6) should not go beyond 0.000001 of the overall thiamin concentration near pH 7 to 8. This concentration drops drastically in basic media where, near pH 11, it should not go beyond the tenth millionth of the overall concentration of thiamin. Therefore, it can be inferred that under all conditions of acidity,[2] the structural transformations of thiamin in neutral and basic media would drastically inhibit the concentration of the catalyst proposed by Breslow. This observation may lead to investigating the catalytic activities of other present intermediates whose intervention would compete with the A_{-H} action and eventually overshadow it.

A. $A^+ \rightleftarrows J^-$ Structural Transformation

If a neutral solution of thiamin (containing A^+) is strongly basified, at least two kinetic phenomena are detected at 339 nm (Figure 5): a sharp rise in absorbance corresponding to the formation of the yellow form J^- (the only thiamin species to have an absorbance spectrum above 300 nm), followed by a slow exponential drop in the concentration of J^- which generates thiolate C^-, the thermodynamic product.

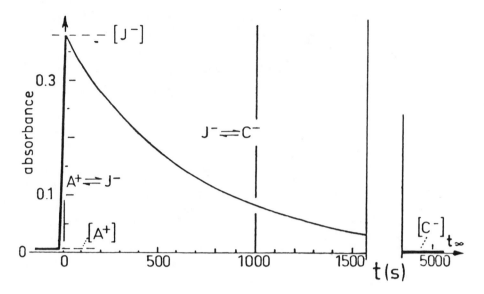

FIGURE 5.   Fast pH jump from neutral (approximate pH 7) to basic (approximate pH 13.26) at 25°C. Absorbance change of a thiamin solution at 339.5 nm (c = 4.90 × 10^{-5} *M*).

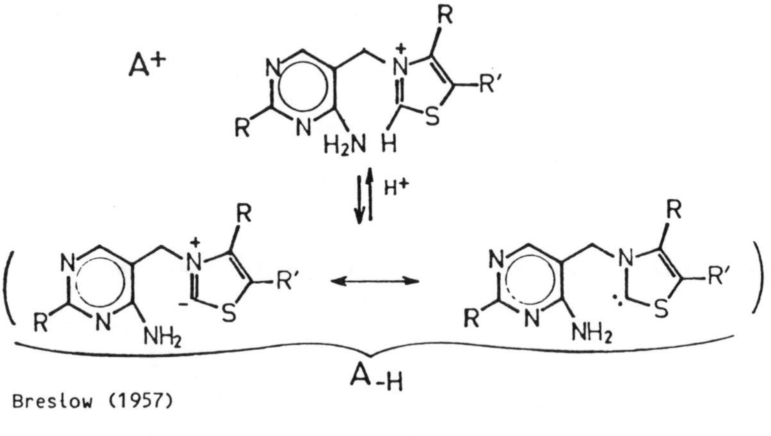

SCHEME II.

The first process is an exponential rise in absorbance studied by stopped-flow kinetics. The fast relaxation mode is the only relaxation mode, as it has only a single relaxation amplitude. Even for the final OH^- concentration of 1 *M*, the formation of J^- detected in the 300 to 400 nm range displays only a single relaxation mode with its associated single relaxation amplitude. This amplitude always corresponds to a variation in absorbance between A^+ (initial species) and J^- (intermediate species) (Figure 5). The deprotonation of A^+ into resonance-stabilized carbene species A_{-H} is extremely fast (a proton transfer), and should be accompanied by a bathochromic shift in absorbance (indicating carbene or zwitterion formation). The high OH^- concentration in some of our experiments clearly indicates that if the formation of the resonance-stabilized species (Scheme II) were involved in the A^+ ⇌ J^- transformation, it would have a pK above 15. With a pK below 15, the relaxation amplitude of thiamin deprotonation would have been detected by stopped-flow kinetics. The formation of resonance-stabilized species A_{-H} with a pK of 12.6 or 14.15[7,8] is not observed.[2] If the deprotonation of cation A^+ (with pK < 15) in the A^+ ⇌ D transformation is taken

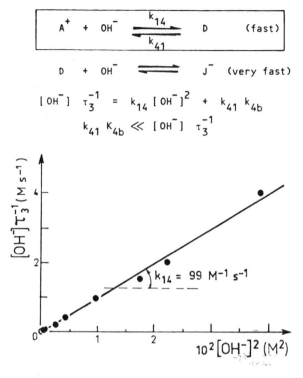

FIGURE 6.   The structural transformation of A^+ into J^- in basic media. Plot of $[OH^-] \tau_3^{-1} = k_{14} [OH^-]^2$.

into account, the experimental data do not fit the reciprocal relaxation equation for the A_{-H} $\rightleftarrows$ D transformation (Scheme II). However, they do fit the reciprocal relaxation equation for the $A^+ \rightleftarrows$ D transformation without A_{-H} involvement (Figure 6).[2] The slope of the best linear least-squares regression equals the k_{14} rate constant of 99 M^{-1} sec^{-1} for intramolecular nucleophilic addition of the amino at thiamin position 2, in good agreement with the value reported by Hopmann et al. for the overall $A^+ \rightleftarrows$ D transformation.[8]

So, cation A^+ changes into yellow form J^- via σ-adduct D probably without any involvement by resonance-stabilized carbene species A_{-H}.[2]

B. $J^- \rightleftarrows C^-$ Structural Transformation

The second phenomenon observed in Figure 5 corresponds to the structural transformation of yellow form J^- into thiolate C^-. According to Hopmann,[7] this transformation would be indirect and semi-irreversible. However, the experimental data do not fit the kinetic equation associated with such a transformation.[2] These same data do fit the reciprocal relaxation equation associated with the indirect transformation of J^- into C^- (via D, A^+, and B) corresponding to the rate-limiting hydroxylation of A^+ into B (Figure 7).[2] The best linear least-squares regression gives 19.2 M^{-1} sec^{-1} for k_{34}, the rate constant for the hydroxylation of cation A^+ into pseudobase B, in good agreement with the rate constant measured for the same transformation in mildly basic media (where J^- does not form) (Figure 4).[1]

These results do not as yet account for the lability of thiamin proton 2 during the transformations of neocyanothiamin into its yellow form reported by Hopmann et al.[8]

VI. LABILITY OF THIAMIN PROTON 2 DURING THE $A^+ \rightleftarrows J^-$ TRANSFORMATION

The proton NMR spectra of cation A^+ and yellow form J^- were recorded in D_2O. In the

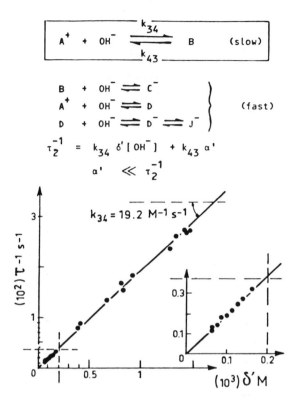

FIGURE 7.   The structural transformation of thiamin in basic media. Plot of $\tau_2^{-1} = k_{34}\,\delta' + k_{43}\,\alpha'$, where α' and δ' are complex functions of the concentrations of the species present in the solution.

experimental medium, J^- was generated by using $NaOD/D_2O$ to strongly basify a slightly acidic D_2O solution of thiamin. The NMR spectrum of J^- was recorded by a single pulse very shortly after this basification.

The spectrum of A^+ recorded in heavy water is not different from the spectrum previously recorded in water by Azahi and Mizuta.[12] However, the newly recorded spectrum of J^- no longer displays a peak at 7.54 ppm (proton 2).[1] The NMR spectrum of A^+ in D_2O (obtained by using DCl to neutralize a solution of J^- to approximately pD 6) no longer displays a peak at 9.83 ppm (proton 2).[2] This neutralization gives rise to an A^+ cation deuterated at position 2.[2] These observations, like those made by Hopmann et al. for neocyanothiamin,[8] tend to indicate the occurrence, during the structural transformation of A^+ into J^-, of an exchange between thiamin proton 2 and the heavy water deuterons. However, as seen above, the $A^+ \rightleftarrows J^-$ transformation does not seem to involve the deprotonation of cation A^+ into A_{-H} which would need a pK > 15. This leaves the possibility that thiamin exchanges its proton 2 during the $D \rightleftarrows J^-$ transformation. This proton exchange probably occurs via carbanion D^- (the deprotonation product of D at position 2) (Figure 8).[2] As site 2 of D is attached to three electronegative atoms (1–O and 2–N), it can probably deprotonate into D^- which, by classical ring opening, can in turn yield J^-.[2]

The involvement of carbanion D^- in the biocatalytic activity of vitamin B_1 is an attractive possibility because it would account for the vital role played by aminopyrimidine in this activity.[2] If the aminopyrimidine group is replaced by any other aromatic ring (e.g., a benzyl or a phenyl), the activity of thiamin in the decarboxylation of pyruvic acid is either totally or partially inhibited.[13] Such inhibition could not be accounted for if zwitterion A_{-H} were the only species of thiamin involved in its activity in metabolism.[2]

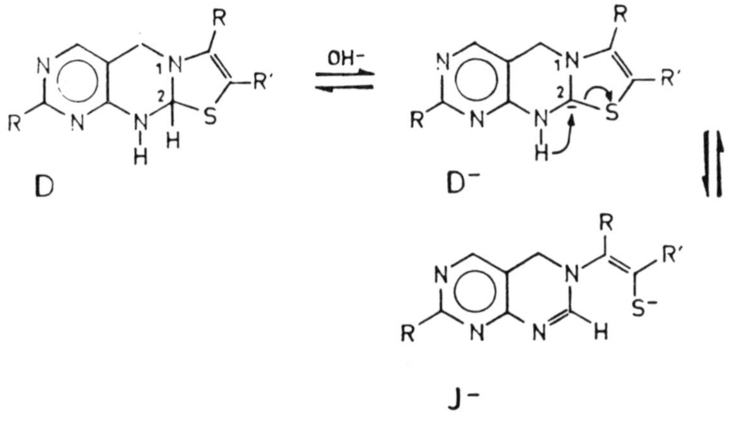

FIGURE 8.   The structural transformation of adduct D into yellow form J⁻ via carbanion D⁻.

VII. CONCLUSION

As a general system, the transformation of thiamin over a broad range of medium conditions reveals a certain number of intermediates involved in its complexity. Maier and Metzler laid the foundation for the overall mechanisms, and we have been able to identify the kinetic aspects of the main transformations. Moreover, we have established the existence of different intermediates in the acid-base transformations that occur at different levels.

It is reasonable to assume two major mechanistic pathways: the first ($A^+ \rightleftarrows C^-$) is under thermodynamic control and the second ($A^+ \rightleftarrows J^-$) is under kinetic control. Both pathways, we feel, could be of biological interest.

A. Thermodynamic Control

In neutral and mildly basic (pH < 11) media, cation A^+ is transformed into pseudobase B by hydration or by rate-limiting hydroxylation. Pseudobase B very rapidly generates thiolate C^- (probably via its deprotonated form B^-),[1] the thermodynamic product in basic media (Scheme III).

B. Kinetic Control

In basic media (pH > 11), thiamin rapidly engenders σ-adduct F, without deprotonation of cation A^+ into resonance-stabilized carbene A_{-H}. D immediately generates J^-, probably via the deprotonation product D (carbanion D^-). J^- in turn slowly yields C^- via D, A^+, and B (which, in basic media, forms by rate-limiting hydroxylation) (Scheme III).

The $A^+ \rightleftarrows J^-$ transformation is under kinetic control. J^- is a kinetic product which evolves into C^- via D, A^+, and B without the accumulation of any intermediate.

C. Biological Interest

At this stage, knowledge of the physical chemistry involved in the transformations of thiamin provides an important set of information bearing on the nature of the intermediate products as well as on their aptitude to evolve. The data acquired make it possible to expand previous hypotheses in order to interpret the in vivo activity of vitamin B_1.

It is very likely that most of the thiamin species and the various reactions involved in their structural transformations play a biological role. To establish their nature was essential, but to identify their role and, therefore, their specific activity in a biological content constitutes another challenge. This is all the more true since even their real concentration can

THIAMIN ENTITIES

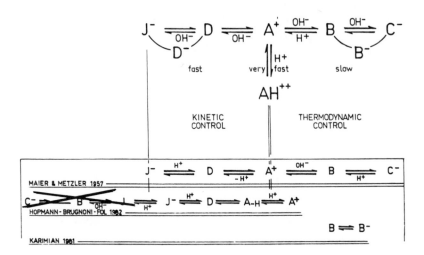

SCHEME III.

be modified according to the micromedia in which thiamin diffusion can take place and evolve according to local pH values. For instance, as thiolate C^- is known to diffuse through the living cell membrane, one could consider that its generation from cation A^+ would be controlled by rate-limiting hydroxylation. In the cell itself (neutral), thiolate C^- would generate A^+ by pseudobase B acid-promoted dehydration, and cation A^+ would immediately enter in equilibrium with D^- or any other catalyst such as A_{-H}. Hence, cation A^+ would play the role of reservoir for D^- or A_{-H}. Thus, the thermodynamic and kinetic aspects of the structural transformations of thiamin would control the concentration of the biocatalyst in the medium and would also monitor the rate of the biocatalysis.

The information, provided thus far by the physical chemistry of thiamin in water, should help to elucidate the bioactivities of some of the intermediates in the living cells, where site binding could displace some equilibria by enhancing the formation of some identified intermediates through their local nucleo- or electrophilic activities. Further investigations following kinetic determinations with relaxation techniques and spectroscopy are in progress. We expect that they will be of major help to further the understanding of the various mechanisms whereby thiamin, by its transformation, acts in living organisms.

REFERENCES

1. **El Hage Chahine, J. M. and Dubois, J. E.**, *J. Am. Chem. Soc.*, 105, 2335, 1983.
2. **El Hage Chahine, J. M. and Dubois, J. E.**, *in press.*
3. **Lehninger, A. L.**, *Biochemistry*, Worth, New York, 1977.
4. **Breslow, R.**, *J. Am. Chem. Soc.*, 80, 3719, 1958.
5. **Olofson, R. A., Thompson, W. R. and Michelman, J. S.**, *J. Am. Chem. Soc.*, 86, 1865, 1964.
6. **Schellenberger, A.**, *Angew. Chem. Int. Engl. Ed.*, 6, 1024, 1967.
7. **Hopmann, R. W.**, *Ann. N. Y. Acad. Sci.*, 378, 32, 1982.
8. **Hopmann, R. W., Brugnoni, G. P., and Fol, B.**, *J. Am. Chem. Soc.*, 104, 1341, 1982.
9. **Dugas, H. and Penney, C.**, *Bioorganic Chemistry*, Springer-Verlag, New York, 1981 (and references cited therein).
10. **Maier, G. D. and Metzler, D. E.**, *J. Am. Chem. Soc.*, 104, 1341, 1982.

11. **Eigen, M. and De Mayer, L.,** *Techniques of Chemistry,* Vol. 6 (Part II), Weissberger, A. and Hammes, G., Eds., John Wiley & Sons, New York, 1973.
11a. **Bernasconi, C. F.,** *Relaxation Kinetics,* Academic Press, New York, 1976.
12. **Azahi, Y. and Mizuta, E.,** *Talanta,* 19, 567, 1972.
13. **Yatco-Manzo, E., Rodd, F., Yount, R. G., and Metzler, D. E.,** *J. Biol. Chem.,* 234, 733, 1959.

Chapter 6

MECHANISM OF NUCLEOPHILIC SUBSTITUTION OF THIAMIN AND ITS ANALOGS

John A. Zoltewicz, Georg Uray, and Ingo Kriessmann

TABLE OF CONTENTS

I. SULFITE ION AS THE NUCLEOPHILE

A. Thiamin and Its Nonquaternized Analogs

Most of the kinetic data pertaining to nucleophilic substitution reactions of vitamin B_1 or thiamin (*1*) (G=H) and its derivatives and analogs refer to the use of sulfite ion as the nucleophile. This is a consequence of the report in 1935 by Williams that *1* is rapidly cleaved by aqueous solutions of this ion. He observed that *1* and sulfite ion give the sulfonic acid *2* and the unquaternized thiazole *3* in a reaction in which sulfite ion becomes attached to the methylene bridge of the vitamin in a nucleophilic substitution process.[1-3]

Not until 1977 was the mechanism of this archetypical substitution reaction of *1* deciphered, owing not to a lack of experimental effort, but to its novel nature and complexity.[4] From the bell-shaped pH rate profile it was apparent that protonated thiamin reacts with unprotonated sulfite ion, but this in itself did not reveal the mechanism. The decisive experiments employed a pair of competing nucleophiles, sulfite and azide ions, the latter chosen because of its long use as a trapping agent for carbocations. The competition experiments revealed that two substitution products form sulfonic acid *2* and a new material, azidomethylpyrimidine *4*. Moreover, and this is the important point, although azide ion became incorporated into the product, it did *not* influence the rate of the substitution reaction. These observations require the presence of a reactive intermediate that reacts with the competing nucleophiles *after* the rate-limiting step, thus eliminating as a possibility the obvious S_N2 mechanism for the observed second-order kinetics. The mechanism of substitution of *1* must be multistep. Moreover, the same observations were made on a thiamin analog *5*, one having pyridine as a leaving group in place of the thiazole ring of *1*. This *5* was two times more reactive than *1*, showing only a modest change in the rate constant in spite of a major modification in structure.[4] Clearly, the pyrimidine ring common to both *1* and *5* must play a vital role in the mechanism, while the nature of the leaving group is of lesser importance.

The most reasonable and economical mechanism consistent with these observations is one in which sulfite ion adds to protonated thiamin at its unsubstituted and therefore most reactive 6' position, the proton being bonded at its thermodynamically favored 1' position. The resultant sigma complex then undergoes fragmentation to liberate the leaving group and give the reactive, resonance-stabilized cationic intermediate *6* (G=H). The driving force for these steps is the conversion of the electron-withdrawing pyrimidine ring into a good electron donor that helps to expel the leaving group and to stabilize the resultant cation by resonance. Intermediate *6* then reacts with nucleophiles such as sulfite and azide ions present in the reaction mixture. Finally, expulsion of sulfite ion regenerates the aromatic pyrimidine ring of *2*. This sequence to sulfonic acid *2* requires the involvement of two sulfite ions, the first being kinetically significant under normal conditions, the second participating after the rate-limiting step.[4]

There is convincing evidence for the involvement of two sulfite ions in the substitution process under special conditions. At very low concentrations of sulfite ion, intermediate *6* competes for this ion and pyrimidine free base, acting now as a nucleophile. Under these atypical reaction conditions, the kinetics become second order in sulfite ion as the multistep mechanism demands.[5,6]

B. Quaternized Thiamin and Its Analogs

The pyrimidine ring of thiamin has to be protonated before reaction with sulfite ion takes place. Obviously, the replacement of a proton by a methyl group at position 1 to give derivative *1* (G=CH$_3$) would provide a substrate with the required activation and thereby simplify kinetic studies; no longer would complete pH rate profiles be required to interpret a mechanism. Therefore, an attempt was made to synthesize the desired derivative. However, N-methylation of *1* as its free base proved to be a challenge because of the facile oligo-

STRUCTURES 1—9.

merization of the pyrimidinylmethyl portion of the product.[7] Thus, in nonaqueous solvents, once a little of the desired N-methylated derivative is formed, it rapidly reacts with starting material in a substitution reaction giving rise to the oligomer. Fortunately, such a process is slow in water and the selection of dimethyl sulfate as the methylating agent makes trivial the synthesis of N-methyl thiamin *1* (G=CH$_3$, N-methyl B$_1$). However, there are as yet no reported good syntheses of other N-quaternized thiamins starting from *1*.

N-methyl B$_1$ is a good starting material for the preparation of a wide variety of N-methylated thiamin analogs. Heating this compound with pyridines, phenoxide ions, or other nucelophiles in methanol gives rise to analogs *7* in which the thiazole ring has been replaced.[8,9]

The results in Table 1 show that such analogs undergo substitution with sulfite ion. Those with phenoxide ion leaving groups (L) demonstrate only a modest rate dependence on the basicity of the leaving group, the slope of a Hammett plot being 1.35.[10] Those with a pyridine leaving group reveal a greater sensitivity to the basicity of the nucleofuge, as seen by the large Bronsted slope of -0.85.[11] The most reactive analog yet reported is that having a 4-

Table 1
RATE CONSTANTS FOR THE REACTION OF
1′-METHYLTHIAMINIUM ION *1* AND ITS
ANALOGS *7* WITH AQUEOUS SULFITE ION
AT 25°C[a]

Leaving group	k_2 (M^{-1} sec^{-1})	Ref.
Thiazole of B$_1$	4.10×10^{-2}	7
Thiazole of B$_1$[b]	1.92×10^{-1}	4
4-Cyanopyridine[c]	19.3	This chapter
3-Carbamoylpyridine	2.4	11
Pyridine	8.71×10^{-2}	11
Pyridine[b]	3.78×10^{-1}	7
4-Methylpyridine	1.10×10^{-2}	11
3,4-Dimethylpyridine	3.49×10^{-3}	11
Azide ion	1.07×10^{-1}	This chapter
4-Nitrophenoxide ion	8.64×10^{-2}	10
4-Cyanophenoxide ion	5.83×10^{-2}	10
3-Chlorophenoxide ion	3.80×10^{-2}	10
3-Methoxyphenoxide ion	1.29×10^{-2}	10
Phenoxide ion	6.36×10^{-3}	10
4-Methylphenoxide ion	5.5×10^{-3}	10
p-Tosylate ion	2.42	12

[a] $\mu = 1$ (KCl).
[b] Unquaternized substrate (*1*, G=H) in D$_2$O.
[c] Followed at 229 nm.
[d] Followed at 290 nm.

cyanopyridine as a leaving group. Remarkably, this substrate is more reactive than that having a tosylate ion-departing entity.[12]

Proof that the 4-cyanopyridine analog and by analogy other pyridine analogs react by the multistep mechanism is found in competition experiments. *p*-Nitrobenzenethiolate ion in the presence of sulfite ion reacts to give *8* (Nuc=SAr) as the major product, but the presence of this thiolate ion has no influence on the rate of disappearance of the starting material, an important observation. This requires a multistep mechanism in which the thiolate ion reacts with an intermediate, likely *6* (G=CH$_3$), after the rate-limiting step. Quantitatively, the thiolate ion is 400 times more reactive than sulfite ion. This value is similar to that of 250 found for another thiolate ion, the ion from 4-thiopyridone, competing with sulfite ion for the same intermediate.[13]

Analog *7* having 4-thiopyridone anion (L) as a leaving group has interesting kinetic properties: in the presence of added leaving groups, rates of substitution are retarded due to a common ion effect. They become second order in sulfite ion, again providing proof of the suggested pathway for sulfite ion in water.[13]

II. METHOXIDE ION AS THE NUCLEOPHILE

A. Thiamin and Nonquaternized Analog

Nucleophilic substitution by the multistep addition-elimination-addition sequence is not restricted to sulfite ion. A similar process takes place in methanol with methoxide ion serving as the kinetically important nucleophile. Thus, when *1* (G=H) or *5* is heated at 71.5°C, ready substitution takes place in the presence of various nucleophiles. Rates of substitution to give *8* (G=H), Table 2, are independent of both the concentration and the identity of the amine nucleophile that becomes incorporated into the isolated product as required by the

Table 2
RATE CONSTANTS FOR THE
REACTION OF 1'-
METHYLTHIAMINIUM ION *1*
AND ITS ANALOGS *7* WITH
METHOXIDE ION IN METHANOL[a]

Leaving group	k_2 (M^{-1} sec^{-1})
Thiazole of B$_1$[b]	1.70×10^{-5}[c]
Pyridine[b]	1.92×10^{-5}[c]
Pyridine[d]	3.15×10^4
4-Methylpyridine[d]	7.78×10^3
3,4-Dimethylpyridine[d]	2.93×10^3
4-Aminopyridine[e]	6.41
4-Nitrophenoxide ion[e]	107
4-Cyanophenoxide ion[e]	45.6
3-Chlorophenoxide ion[e]	9.12
Phenoxide ion[e]	1.07

[a] Reference 15.
[b] Unquaternized substrate (*1*, G=H); Reference 14.
[c] Pseudo-first-order constant at 71.5°C.
[d] At 25°C.
[e] At 40°C.

mechanism. There is no observed kinetic dependence on methoxide ion because protonated substrate reacts with this ion, kinetically a pH-independent process overall.[14]

B. Quaternized Thiamin and Its Analogs

Rates of substitution become first order in methoxide ion when the substrate is N-methylated, avoiding the need for prior protonation. Again, leaving groups (L) are phenoxide ions and pyridines (Table 2). The nucleophiles are amines present as the buffer. In spite of the fact that these amines are present at some 10^6 times higher concentrations than methoxide ion, it is the latter ion that reacts with substrate according to the observed kinetics. However, the amine and not methoxide ion is found in the major product, an observation consistent with the proposed multistep pathway.[15] On a preparative scale starting with N-methyl B$_1$ analogs *8* may be prepared conveniently in good yield in this way.[9,15]

Second-order rate constants for the pyridine analogs *7* reacting to form *8* show a marked dependence on the basicity of the pyridine. For example, that analog having 4-aminopyridine as a leaving group is some 5×10^5 times less reactive than that with pyridine as the nucleofuge. By contrast, phenoxide ion departs only 100 times slower than *p*-nitrophenoxide ion.[15]

III. HYDROXIDE ION AS THE NUCLEOPHILE

A. Quaternized Thiamin Analogs

Thiamin does not undergo rapid substitution in aqueous alkali because the thiazolium ring cleaves and thereby converts this potential leaving group into an anionic derivative that deactivates the material for substitution.[16] However, quaternized analogs *7* having alkali-stable leaving groups such as pyridine or, better, phenoxide ions do react with hydroxide ion to give the expected substitution compounds *8* (Table 3). That the mechanism again is multistep was demonstrated for those analogs having pyridine or 4-cyanophenoxide ion as

Table 3
RATE CONSTANTS FOR THE
REACTION OF THIAMIN
ANALOGS 7 WITH HYDROXIDE
ION AT 25°C[a]

Leaving group	k_2 (M^{-1} sec^{-1})
Pyridine[b]	2.69
4-Nitrophenoxide ion	4.63×10^{-1}
4-Cyanophenoxide ion	3.00×10^{-1}
3-Chlorophenoxide ion	2.10×10^{-1}
4-Chlorophenoxide ion	1.40×10^{-1}
3-Methoxyphenoxide ion	1.00×10^{-1}
Phenoxide ion	5.61×10^{-2}

[a] $\mu = 1$ (KCl); Reference 15.
[b]   Reference 17.

a departing group. *p*-Nitrobenzenethiolate ion reacted with intermediate 9 (R=H) formed from either substrate to give a major amount of substitution product without influencing the rates of substitution.[17]

A Hammett plot may be constructed for the phenol substrates with a rho value of only 1.08,[15] one similar to that of 1.35 for some of the same substrates reacting with sulfite ion.[10]

IV. COMPARISON OF REACTIVITIES

The two families of leaving groups, pyridines and phenoxide ions, were examined with the three nucleophiles, sulfite, methoxide, and hydroxide ions. The study involving pyridines and hydroxide ions is incomplete as yet. In all cases, the same reactivity pattern was found within a family for a given nucleophile as required for a common mechanism of substitution.

Substitution on pyridine substrates is remarkably slower in water than in methanol. For example, the rate constant for the compound with pyridine as a leaving group is 1.2×10^4 times smaller when hydroxide ion is the nucleophile than when it is methoxide ion.[15] Perhaps the surprisingly high reactivity of the pyridine substrates in methanol is due to the presence of two positive charges which are less well solvated in the less polar methanol than in water and thereby electrostatically destabilize the ground state, fewer charges being present in the transition state.

While the best-known nucleophile for substitution in the thiamin series is sulfite ion, this ion is not the most reactive. Comparing the reactivities of a single substrate toward the three nucleophiles is revealing. Toward 7 having pyridine as a leaving group, e.g., the relative order for sulfite, hydroxide, and methoxide ions, respectively, is $1:31:3.6 \times 10^5$, and for 7 when phenoxide ion departs this same reactivity order is 1:8.8:170 (Tables 1 to 3). The unique feature of sulfite ion is that it reacts even at low pH, a reflection of its relatively low basicity.

REFERENCES

1. **Williams, R. R.**, Structure of vitamin B, *J. Am. Chem. Soc.*, 57, 229, 1935.
2. **Williams, R. R., Buchman, E. R., and Ruehle, A. E.**, Studies of crystalline vitamin B. VIII. Sulfite cleavage. II. Chemistry of the acidic product, *J. Am. Chem. Soc.*, 57, 1093, 1935.

3. **Williams, R. R., Waterman, R. E., Keresztesy, J. C.,and Buchman, E. R.,** Studies of crystalline vitamin B. III. Cleavage of vitamin with sulfite, *J. Am. Chem. Soc.,* 57, 536, 1935.
4. **Zoltewicz, J. A. and Kauffman, G. M.,** Kinetics and mechanism of the cleavage of thiamin, 2-(1-hydroxyethyl) thiamin and a derivative by bisulfite ion in aqueous solution. Evidence for an intermediate, *J. Am. Chem. Soc.,* 99, 3134, 1977.
5. **Doerge, D. A. and Ingraham, L. L.,** Kinetics of thiamine cleavage by bisulfite ion, *J. Am. Chem. Soc.,* 102, 4828, 1980.
6. **Zoltewicz, J. A. and Uray, G.,** Nature of the reaction of thiamin in the presence of low concentrations of sulfite ion. Competitive trapping, *J. Org. Chem.,* 46, 2398, 1981.
7. **Zoltewicz, J. A., Uray, G., and Kauffman, G. M.,** Preparation and reactivity of model compounds related to oligomers from thiamin. A mechanism for oligomerization, *J. Am. Chem. Soc.,* 103, 4900, 1981.
8. **Zoltewicz, J. A. and Baugh, T. D.,** An improved synthesis of 1'-methylthiaminium salts, *Synthesis,* 217, 1980.
9. **Zoltewicz, J. A.,** Facile nucleophilic substitution reactions of 1'-methylthiaminium salts, *Synthesis,* 218, 1980.
10. **Zoltewicz, J. A., Kauffman, G. M., and Uray, G.,** A mechanism for sulphite ion reacting with vitamin B_1 and its analogues, *Food Chem.,* 15, 75, 1984.
11. **Zoltewicz, J. A., Kauffman, G., and Uray, G.,** Nucleophilic substitution reactions of thiamin and its derivatives, *Ann. N.Y. Acad. Sci.,* 378, 7, 1982.
12. **Uray, G., Kriessmann, I., and Zoltewicz, J. A.,** Nucleophilic substitution by sulfite ion on a thiamin analogue having a good leaving group, *J. Org. Chem.,* 49, 5018, 1984.
13. **Zoltewicz, J. A., Uray, G., and Kauffman, G. M.,** Evidence for an intermediate in nucleophilic substitution of a thiamin analogue. Change from first- to second-order kinetics in sulfite ion, *J. Am. Chem. Soc.,* 102, 3653, 1980.
14. **Zoltewicz, J. A. and Baugh, T. D.,** A multistep mechanism of nucleophilic substitution of vitamin B_1 in methanol, *Bioorg. Chem.,* 13, 209, 1985.
15. **Zoltewicz, J. A., Uray, G., Baugh, T. D., and Schultz, H.,** Mechanism of nucleophilic substitution of thiamine and its analogues. Methanol and water solvents, *Bioorg. Chem.,* 13, 135, 1985.
16. **Maier, G. C. and Metzler, D. E.,** Structures of thiamin in basic solution, *J. Am. Chem. Soc.,* 79, 4386, 1957.
17. **Zoltewicz, J. A. and Uray, G.,** Hydroxide ion as a catalyst for nucleophilic substitution of thiamin analogues by thiolate ions. A rival for sulfite ion, *J. Am. Chem. Soc.,* 103, 683, 1981.

Chapter 7

TRIPHOSPHATES OF THIAMIN DERIVATIVES AS PRECURSORS OF THE COENZYME OR ITS ANTIMETABOLITES

Yu.M. Ostrovsky

TABLE OF CONTENTS

I. INTRODUCTION

Accessibility and concentration of thiamin diphosphate (ThPP) in the cell or its organelles to a very large extent determine functioning of alphaketo acid dehydrogenases and transketolase. Animal tissues always have certain stores of nonproteinized ThPP with the extremely labile level which changes toward a drastic decrease or elevation in thiamin deficiency or after supplementary administration of the vitamin.[1,2] Under normal conditions, free thiamin, formed from thiamin phosphates (ThP, ThPP, ThPPP) as a result of the action of unspecific phosphatases in the lumen of the intestines, mainly enters blood from the gastrointestinal tract of animals and humans. However, tissues, blood, and cerebrospinal fluid are rich in thiamin monophosphate (ThP) and thiamin triphosphate (ThPPP), and their significance in metabolism of the vitamin itself has been studied very little. In addition to being of purely theoretical interest, this question is important due to the fact that ThPP is applied extensively as a drug empirically, without sufficient scientific substantiation. It is suggested that parenterally administered ThPP directly achieves its object, occupying a position as coenzyme on the respective apoproteins. In this situation, holotransketolase is produced when ThPP overcomes only the cell barrier because the enzyme is located in the hyaloplasm,[3] while to form alphaketo acid dehydrogenases, the coenzyme must penetrate into mitochondria. As regards the latter, the localization of thiamin pyrophosphokinase (EC 2.7.6.2) in the cytosol and the lack of it in mitochondria,[4,5] i.e., the absence of independent synthesis of the coenzyme in the organelles, make the situation even more problematic.

In this laboratory, the last 5 years were devoted to detailed studies on the mechanisms of ThPP transport into the animal organism and the possible role of ThPPP and its analogs as precursors of the coenzyme or anticoenzyme.

II. PERMEABILITY OF ThPP INTO RAT HEPATOCYTES

We previously showed[1] that dilution of the label in different tissues is equal (within 1 to 24 hr) when ^{35}S-thiamin is administered in combination with a 10-fold amount of the unlabeled vitamin or ThPP. The results obtained may be interpreted easily if we admit that after the administration into the body, ThPP immediately undergoes complete dephosphorylation. We did not also observe a significant, compared to the free vitamin, increase in the tissue coenzyme level after ThPP injections.

Direct experiments with double label (^{14}C and ^{33}P) could give a definite answer about the mechanism of coenzyme penetration into the cell. To this end, a highly active preparation of thiamin pyrophosphokinase was used in our laboratory[6] to synthesize ThP^{33}P and 2-^{14}C-ThP^{33}P from thiamin, 2-^{14}C-thiamin, and γ-^{33}P-ATP. These compounds were administered into the portal vein of hexenal-anesthetized rats. In 10 min, livers were excised. Liver preparations were assayed for content of the labels in fractions of ThPP and nucleoside triphosphates after denaturation of proteins and separation on a SP-Sephadex G-25 column. In control experiments, animals were injected with an equivalent amount of H_3^{33}PO$_4$. Under these conditions, almost all the radioactive phosphorus turned out to be in the nucleoside triphosphate fraction, and only 3 or 4% of it incorporated into ThPP after the administration of both ThP^{33}P and ^{33}P$_i$. The results obtained suggest essentially complete dephosphorylation of ThPP administered into blood on passing across histohematic barriers. The experiment was finally confirmed by the results of 2-^{14}C-ThP^{33}P administration into blood (the ratio of the labels was 2:1). The chromatographic separation of the extract showed that the total label was distributed in the following manner: 76% was within thiamin and ThP, 17% was within nucleotides, and ThPP contained 7% of it. The nucleotide fraction showed only ^{33}P. No labeled carbon was found. Practically the whole label incorporated into ThPP as ^{14}C. The coenzyme contained only trace amounts of ^{33}P since there was less ^{33}P in the initial preparation than in the previous experiment with ThP^{33}P and H$_3$^{33}PO$_4$. Accordingly, reutilization of inorganic phosphate via oxidative phosphorylation and the thiamin pyrophosphokinase reaction turned out to be on the limit of sensitivity of the method.

All the data cited only show that prior to getting into the hepatocyte, the ThPP administered into blood has been completely dephosphorylated. Using another technique, which evaluated permeability of ^{14}C-thiamin, ^{14}C-ThP, and ^{14}C-ThPP into hepatocytes and their competition with the unlabeled vitamin, 5-Me-Th, ThP, ThPP or pyrithiamin, other authors[7] arrived at the same conclusion simultaneously with us.

III. ThPPP AS PRECURSOR OF ThPP

As early as the 1950s, ThPPP was discovered in yeast and rat liver.[8,9] These data were later confirmed by other authors who found ThPPP in yeast, plants, and different animal tissues.[10-16] With rare exception,[13] all the researchers established that ThPPP amounted to 1 to 10% of the total content of the vitamin in the object. Even earlier works considered the possible functions of ThPPP as a coenzyme precursor,[17] a participant in reversible exchange of high-energy phosphates,[18,19] or ion transport in the neuron.[20]

Naturally, the search for specific enzyme systems responsible for ThPPP metabolism in the cell started at once. ThPPP formation was first demonstrated in yeast as a result of the transfer of the ATP terminal phosphate to ThPP.[21] Partial purification of the enzyme from nervous tissue (ATP:ThPP) phosphotransferase, EC 2.4.7.15) was first performed in 1964[22] and then in 1968.[23] It was shown later[24] that the true substrate for the liver enzyme is a ThPP-protein complex. The enzyme from liver tissue and yeast was investigated in this laboratory.

The enzyme from the rat liver hyaloplasm has been purified 70-fold[25] to homogeneity on polyacrylamide gel electrophoresis. The protein is a dimer with a molecular weight of 56,000; it has two pH optima in acid (5.0 to 6.5) and slightly alkaline (7.0 to 8.5) media. The reaction rate vs. ThPP concentration curve is S-shaped with the plateau at a concentration of $2.5 \cdot 10^{-5}$ M. The specific activity of the enzyme is rather low (1 nmol).

The yeast enzyme purified 2000-fold[26] is a tetramer with a molecular weight of 90,000. The curves of dependence of the enzyme-specific activity on the substrates (ThPP, Mg·ATP) and the effector (Mg^{2+}) are sigmoidal in form. The protein belongs to the class of dissociating allosteric enzymes.

It is to be noted that the only substrate for both enzyme preparations was free ThPP — not a protein-ThPP complex as was earlier assumed for a similar enzyme.[24]

ThPPP hydrolysis in the cytosol fraction of brain homogenate[27] and the membrane fraction isolated from the liver[28] were simultaneously shown by different authors. Specificity of the ThPPP hydrolysis, however, might easily be disputed due to the ability of widespread alkaline and acid phosphatases[29] to hydrolyze different thiamin phosphates. Separation of the phosphatase activities in the cytosol fractions of liver, brain, kidney, heart, skeletal muscles, spleen, and small intestine homogenates by means of gel electrophoresis and Sephadex G-100 fractionation made it possible to differentiate clearly between various enzymes.[16]

A protein specific to ThPPP (EC 3.6.1.28) was found in all tissues with the exception of the intestines. In the latter, ThPPP hydrolysis was performed by means of alkaline and acid phosphatases. ThPPP-ase has low molecular weight (30,000) and Stokes radius of around 2.5 nm. The isoelectric point (pI) for the brain enzyme is 4.6 at 4°C. Being partially purified by ammonium sulfate fractionation, brain ThPPP-ase is absolutely inactive (at pH 9.0) with other thiamin phosphates, nucleoside triphosphates, and *p*-nitrophenyl phosphate. The liver enzyme is four- to sixfold less active, compared to ThPPP, with CTP, UTP, ITP, and GTP, but not with adenylic nucleotides. Since the enzyme has not yet been purified to homogeneity, these data should also be accepted with a proviso. The apparent Michaelis constant for

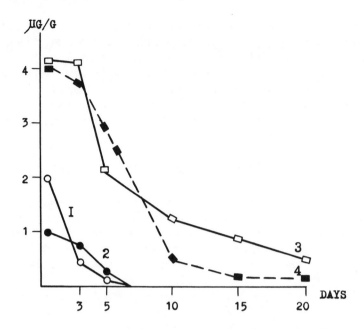

FIGURE 1.   Content of thiamin (1), ThP (2), ThPP (3), and ThPPP(4) in the rat liver over different periods of vitamin B₁ deficiency.

partially purified ThPPP-ase from rat brain and liver is 0.5 m*M*; the specific activity of the liver enzyme is 3 or 4 orders higher[16] than the rate of ThPPP synthesis.[25]

In our studies, we took an interest in ThPPP as an active component of the vitamin pool. In this context, we revised the data on tissue level of this derivative and determined efficiency of ThPPP as a coenzyme precursor.

On determination of tissue ThPPP level under conditions of the maximally rapid inactivation of ThPPP-ase, we always found approximately equal amounts of ThPPP and ThPP.[30] In thiamin deficiency, Th and ThP left the liver very rapidly, while ThPP and ThPPP disappeared at nearly the same rate (Figure 1). When animals were injected singly with thiamin on the 20th day of deficiency, ThPP was produced more rapidly than ThPPP, but the ThPPP concentration still increased even at decreasing levels of the coenzyme (Figure 2).

When ThPPP is administered parenterally into animals, the tissue ThPP content is drastically increased over the shortest periods. The coenzyme level in mitochondria obtained from the rat liver on the 5th and 10th days of vitamin B₁ deficiency cannot be increased pronouncedly, incubating the organelles with ThPP for 10 min, but it can be doubled by adding ThPPP to the incubation medium.[31] Higher compared to other thiamin derivatives, vitamin activity of ThPPP also attracted the attention of scientists 20 years ago.[32]

IV. TRIPHOSPHATES OF THIAMIN DERIVATIVES AS ANTICOENZYME PRECURSORS

The studies on tetrahydrothiamin and thiochrome diphosphate esters as anticoenzymes in ThPP-dependent reactions were carried out fairly long ago.[33] The testing of the antivitamin activities of the same compounds or their precursors in experiments on animals yielded absolutely negative results.[34-36] The data obtained are in full agreement with our thesis about the impermeability of ThPP through histohematic barriers,[6] and are quite explicable if we take into account that neither tetrahydrothiamin nor thiochrome is phosphorylated by thiamin

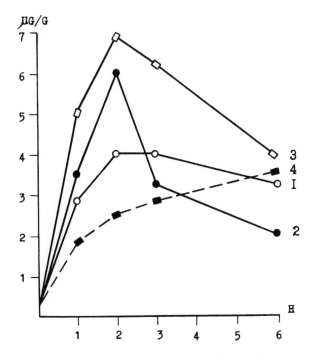

FIGURE 2. Content of thiamin derivatives in the liver (designations are the same as in Figure 1) at different time intervals after administration of 2 mg/kg thiamin into animals on the 20th day of deficiency.

Table 1
PYRUVATE DEHYDROGENASE ACTIVITY
(μmol/g protein/20 min) 3 HR AFTER
ADMINISTRATION OF DIFFERENT THIAMIN
DERIVATIVES INTO MICE

Derivatives	Dose (mg/kg)	Liver	Heart
Control	—	122 ± 4	216 ± 8
Hydroxythiamin	10	121 ± 3	215 ± 4
Pyrithiamin	10	130 ± 7	230 ± 9
Tetrahydro-ThPPP	10	50 ± 6[a]	120 ± 11[a]
Thiochrome-PPP	10	47 ± 2[a]	134 ± 5[a]
	200	40 ± 2[a]	95 ± 4[a]

[a] $p > 0.01$ to 0.001.

pyrophosphokinase to diphosphates.[37] As early as 3 hr after administration of triphosphate esters of the same thiamin derivatives, the activities of liver and heart pyruvate dehydrogenase[36] were sharply depressed (Table 1). In this experimental variant, the classic thiamin anti-metabolites, pyrithiamin and hydroxythiamin, had not had time to manifest their antivitamin properties.

On the basis of the total combination of the data on ThPPP and its analogs, we may admit that the triphosphates of thiamin and its derivatives penetrate through histohematic barriers, removing the terminal phosphate. As a result, highly active diphosphate esters turn out to be the other side of the membrane. The mechanism remains that known for transport of

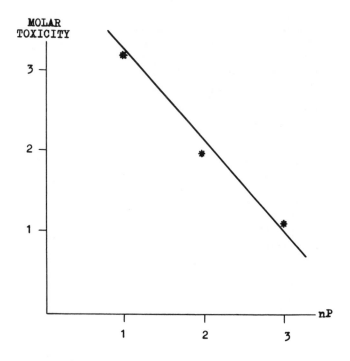

FIGURE 3.   Molar toxicity (LD_{50}/mol wt) of tetrahydrothiamin-phosphorylated derivatives.

adenylic nucleotides in the cell.[38] Two independent mechanisms of this process for ThPPP possibly occur: energy independent, involving ThPPP-ase, and more interesting from the regulatory viewpoints, interaction of ThPPP with ADP. The second probability was demonstrated long ago in studying the fate of the terminal phosphate of ThPPP in rephosphorylation reactions.[18,19] It may not be out of place to emphasize that only the combination of ThPPP and ADP completely reactivates oxoglutarate dehydrogenase of yeast grown in thiamin-free medium or the mitochondrial enzyme from deficient animals.[17] It is also interesting that partial reactivation of the enzyme by means of ThPP was accompanied by weak binding of the coenzyme (splitting off in alkaline medium), whereas reactivation in the presence of ThPPP and ADP resulted in tight, probably covalent binding of the coenzyme.

High biological activity of ThPPP was clearly elucidated by the pharmacological study.[40] With ThPPP analogs, we obtained similar results when determining molar toxicity (LD_{50}/mol wt) (Figure 3); tetrahydro-ThPPP turned out to be more toxic in the homologous series.

V. CONCLUSION

In the total pool of thiamin, ThPPP is a natural component presumably fulfilling the functions of depot and coenzyme precursor. Triphosphate esters of thiamin analogs are proposed as compounds for rapid and efficient inhibition of the activities of pyruvate dehydrogenase in experiments in vivo.

REFERENCES

1. **Ostrovsky, Yu.M.**, Active centers and groups in the thiamine molecule, *Nauk. Tekh.*, p. 422, 1975.
2. **Ostrovsky, Yu.M., Voskoboyev, A. I., Gritsenko, E. A., and Grushnik, V. V.**, *Prikl. Biokhim. Mikrobiol.*, 15, 728, 1979.
3. **Voskoboyev, A. I., Averin, V. A., and Ostrovsky, Yu.M.**, *Vopr. Med. Khim.*, 27(3), 366, 1981.
4. **Deus, B. and Blum, H.**, *Biochim. Biophys. Acta*, 219, 489, 1970.
5. **Voskoboyev, A. I. and Averin, V. A.**, *Vopr. Med. Khim.*, 27(2), 239, 1981.
6. **Voskoboyev, A. I. and Ostrovsky, Yu.M.**, *Vopr. Med. Khim.*, 29(4), 42, 1983.
7. **Yoshioka, K., Nishimura, H., Sempuku, K., and Iwashima, A.**, *Experientia*, 39(5), 505, 1983.
8. **Kiessling, K.-H.**, *Nature (London)*, 172(4391), 1187, 1953.
9. **Rossi-Fanelli, A., Siliprandi, N., and Fasella, P.**, *Science*, 116, 711, 1952.
10. **Yusa, T.**, *Plant Cell Physiol.*, 2(4), 471, 1961.
11. **Rindi, G. and Giuseppe, L.**, *Acta Vitaminol.*, 14(6), 245, 1960.
12. **Iida, S.**, *Biochem. Pharmacol.*, 15(8), 1139, 1966.
13. **Eder, L. and Yves, D.**, *J. Neurochem.*, 35(6), 1278, 1980.
14. **Matsuda, T. and Cooper, J. R.**, *Anal. Biochem.*, 117(1), 203, 1981.
15. **Rindi, G.**, *Acta Vitaminol. Enzymol.*, 4(1 and 2), 59, 1282.
16. **Penttinen, H. K. and Uotila, L.**, *Med. Biol.*, 59, 177, 1981.
17. **Yusa, T. and Maruo, B.**, *J. Biochem.*, 60(6), 735, 1966.
18. **Greiling, H. and Kisow, L.**, *Z. Naturforsch.*, 116, 491, 1956.
19. **Yusa, T.**, *Plant Cell Physiol.*, 3(1), 95, 1962.
20. **Hoffmann, H., Eckert, T., and Möbus, W.**, *Z. Physiol. Chem.*, 335, 156, 1964.
21. **Kiessling, K.-H.**, *Acta Chem. Scand.*, 13(7), 1358, 1959.
22. **Eckert, T. and Möbus, W.**, *Z. Physiol. Chem.*, 338, 286, 1964.
23. **Hokawa, Y. and Cooper, J.**, *Biochim. Biophys. Acta*, 158(1), 180, 1968.
24. **Ruenwongsa, P. and Cooper, J.**, *Biochim. Biophys. Acta*, 482, 64, 1977.
25. **Voskoboyev, A. I. and Luchko, V. S.**, *Vopr. Med. Khim.*, 26(4), 564, 1980.
26. **Chernikevich, I. P., Luchko, V. S., Voskoboyev, A. I., and Ostrovsky, Yu.M.**, *Biokhimiya*, 49(6), 899, 1984.
27. **Hashitani, Y. and Cooper, J.**, *J. Biol. Chem.*, 247, 2117, 1972.
28. **Barchi, R. I. and Braun, P. E.**, *J. Biol. Chem.*, 247, 7668, 1972.
29. **Kiessling, K.-H.**, *Biochim. Biophys. Acta*, 43, 335, 1960.
30. **Voskoboyev, A. I., Gritsenko, E. A., and Ostrovsky, Yu.M.**, *Vopr. Med. Khim.*, 30(4), 106, 1984.
31. **Averin, V. A.**, Transport of Thiamine across Biological Membranes, Candidate thesis, Vilnius, 1983.
32. **Rosanov, A. Ya.**, Metabolism of Thiamine: Its Phosphate Esters and Disulfides in the Animal Organism, Doctoral thesis, Odessa State University, Odessa, 1963.
33. **Wittorf, J. H.**, Doctoral thesis, Brigham Young University, Provo, Utah, 1968.
34. **Ostrovsky, Yu.M. and Zabrodskaya, S. V.**, *Dokl. Akad. Nauk BSSR*, 24(6), 558, 1980.
35. **Ostrovsky, Yu.M., Zabrodskaya, S. V., Zimatkina, T. I., and Oparin, D. A.**, *Biokhimiya*, 48(6), 928, 1983.
36. **Ostrovsky, Yu.M., Oparin, D. A., Zimatkina, T. I., et al.**, *Izv. Acad. Nauk BSSR. Ser. Biol.*, 6, 102, 1982.
37. **Ostrovsky, Yu.M., Chernikevich, I. P., Voskoboyev, A. I., and Schellenberger, A.**, *Bioorg. Khim.*, 3(8), 1083, 1977.
38. **Reynafarye, B. and Lehninger, A. L.**, *Proc. Natl. Acad. Sci. U.S.A.*, 75, 4788, 1978.
39. **Yamamoto, H.**, *Folia Pharm.*, 63(3), 134, 1967.
40. **Zabrodskaya, S. V.**, Tetrahydrothiamine: Its Phosphorylated Derivatives and Their Effect on Activities of Thiamine-Dependent Enzymes in Animal Tissues, Candidate thesis, Minsk, 1984.

Part II
Pyruvate Decarboxylase: Structure and Mechanism of Action

Chapter 8

PYRUVATE DECARBOXYLASE FROM YEAST AND WHEAT GERM: STRUCTURAL SIMILARITIES AND DIFFERENCES

Johannes Ullrich and Hartmut Zehender

Pyruvate decarboxylase (PDC) (EC 4.1.1.1) isolated from yeast[1-3] tends to be partially degraded by proteinases carried along through all purification steps[4] which incidently become active by mutual cleavage of their specific inhibitors.[5] A limited number of breaks within the chains and even losses of small pieces of chains do not seem to interfere seriously with the enzymatic properties of PDC and hence with enzymological and chemical modification studies.[6] Chemical analysis of the protein, however, requires strictly homogenous polypeptide chains, and this condition has not been met by the PDC preparations hitherto obtained from yeast. Therefore, we included wheat germ PDC in our investigations, hoping to avoid proteolytic or other alterations of the enzyme originally contained in the starting material.

PDC from locally available wheat germs, purified by the classical method[7] and tested by dodecylsulfate polyacrylamide disc electrophoresis,[8] was found to be rather impure. Thus, we had to modify and extend the procedure and succeeded in obtaining homogenous PDC by the following sequence of steps: (1) acetone powder;[7] (2) crude extract;[7] (3) streptomycin precipitation of inactive material; (4) pH 5.2 precipitation of inactive proteins; (5) pH 4.6 precipitation of PDC; (6) gel filtration through Sephacryl S-300 superfine; (7) anion-exchange chromatography over DEAE-Sephacel; and (8) native electrophoresis on polyacrylamide gel gradients[9] as the inevitable final step. Published methods for PDC detection in electrophoretic gels were found to be slow and unsatisfactory[2] or not applicable at all.[10] Therefore, we had to develop a rapid and sufficiently sensitive method for PDC activity staining,[11] based on the precipitation of the insoluble cyclic condensation product, 1,3-diphenyl-2-methylimidazolidine, of the enzymatically formed acetaldehyde with added 1,2-dianilinoethane.[12]

Two bands of PDC activity appeared in electrophoreses of crude wheat germ PDC (see Reference 11). Whereas the major band (LM-PDC, described below) showed a high similarity to yeast PDC, the minor band (HM-PDC) had a considerably higher $M_r = 276,000$ (in electrophoreses) and consisted — to our surprise — of only one sort of chains with $M_r = 33,000$. Comparisons with earlier values of the native molecular weight of wheat germ PDC [$M_r > 10^6$ (Reference 7) and $M_r = 570,000$ (Reference 13)] suggest the possible existence of multiples of 276,000, but this remains unproven, since no data are available on the chain length in these early preparations. Having only very small quantities of HM-PDC, we did not investigate it further and concentrated our efforts on the major fraction.

LM-PDC could be extracted from superficially stained gel bands with appropriate buffers in the native state. Its specific activity under cofactor saturation at pH 6.0 and 30° was approximately 50 U/mg, which is approximately half that of the best yeast PDC preparations obtained as of yet. In native polyacrylamide gel gradient electrophoreses, $M_r \approx 180,000$ (yeast PDC under the same conditions: $M_r \approx 220,000$). After electrophoretic isolation, it was free of cofactors and therefore assumed to exist in a rapid dissociation-association equilibrium like yeast apo-PDC.[2,14,15] However, in gel filtration through Sepharose 6B or Sephacryl S-300 superfine, in the presence or absence of cofactors, an apparent $M_r \approx 390,000$ was found, which was too high for the tetrameric structure expected from extrapolations of the behavior of yeast PDC. As in yeast PDC,[4,15,16] two sorts of polypeptide chains were found in sodium dodecylsulfate disc electrophoresis: α: $M_r = 65,000$ and β: $M_r = 61,000$. We determined the chain weights of yeast PDC under exactly the same conditions: α: $M_r = 63,000$ ($62,000^{16}$) and β: $M_r = 61,000$ ($59,000^{16}$).

LM-PDC was found to decarboxylate pyruvate with a K_m = 3.0 mM (3.6 mM;[7] yeast PDC: K_m = 1.0 mM). It could be shown to exhibit the same cooperative behavior as yeast PDC (n_H = 1.8 to 1.9) and a similar lag phase preceding the start of the enzymatic reaction, particularly noticeable at low substrate concentrations, as had been found for yeast PDC.[17] Pyruvamide, a nonconvertible analog of pyruvate, was found to abolish the lag phase and the cooperativity, demonstrating the existence of activation sites on this PDC as well.

The rapid inactivation of wheat germ PDC by SH reagents[7] could be confirmed. Its sensitivity to oxidation by air was found to be far greater than that of yeast PDC. Thus, during purification, steady SH protection had to be maintained, whereas yeast PDC can be purified in the absence of protective SH compounds without much loss of activity. Apparently, the frequent shelf deterioration of wheat germ PDC is due mainly to oxidation, while yeast PDC under the same conditions is degraded mainly by proteolysis.

We had little success with attempts at determining the pI of wheat germ PDC by a gel electrophoretic method,[18] applying the electric field perpendicular to a pH gradient previously formed on the agarose gel which contains the enzyme in a central groove parallel to the pH gradient. Only a very uneven and diffuse protein distribution could be found — 1.5 pH units up and down from the expected pI — apparently due to precipitation and very low solubility in this range, whereas both wings of the curve and the entire curves of the applied reference proteins were completely sharp. This effect was found for wheat germ as well as for yeast PDCs and may explain the rather variable pI values for yeast PDC reported in the literature as yet and our inability to use chromatofocusing in the course of purification of both PDCs.

Tests for lipophilicity of the active sites of wheat germ PDC with 2-*p*-toluidinonaphthalene-6-sulfonate (TNS) revealed a somewhat lower fluorescence intensity of enzyme-bound TNS than with yeast PDC, although its inhibition of pyruvate binding was the same within the limits of error (K_i = 0.04 mM).

Modification of wheat germ PDC with 2-hydroxy-5-nitrobenzyl-dimethylsulfonium bromide (tryptophan specific under SH protection) gave essentially the same results as with yeast PDC,[6,21] namely, total and irreversible loss of enzymatic activity and coenzyme binding capacity when applied to apo-PDC, and no dramatic change when applied to holo-PDC. Quantitation of the modified residues was less exact than with yeast PDC, due to less tight binding of cofactors and their lower protecting efficiency in holo-PDC.

Since only very small amounts of material were available, no proof could be given by optical methods for electronic interaction and excitation energy transfer between the essential tryptophans and bound thiamin diphosphate or TNS. Nevertheless, the essential tryptophan residues seem to play the same role in wheat germ PDC as in the yeast enzyme.

The notorious contamination of preparations of yeast PDC by proteinases[4,6] and the often lower dyeing intensity of the β band on our analytic gels had induced the assumption that the slightly shorter β chain of PDC may be formed from the α chain by limited proteolysis, either physiologically or as a preparation artifact. This hypothesis had received further support by the detection of only one structural gene coding for PDC in a haploid yeast strain.[22,23] A second PDC gene appeared to be related more to the regulation than to the structure of the enzyme, but showed partial cross-hybridization with the structural gene. Its nature and significance remain obscure. The length of the well-characterized structural gene (approximately 2000 bp) corresponds favorably with that of either PDC chain (530 to 550 amino acid residues). Proof or disproof of the hypothesis could be expected by some chemical analyses.

Tests for carbohydrate were negative for either chain in wheat germ and yeast PDC. So the difference must lie in the number and composition of amino acid residues of the polypeptide chains. The N termini of both chains of wheat germ PDC, as found by the dansyl method, were identical: threonin.

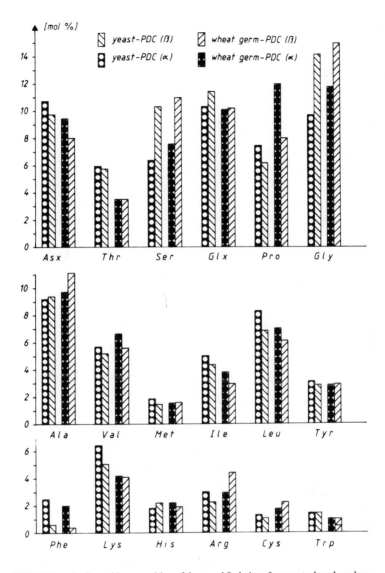

FIGURE 1. Amino acid composition of the α and β chains of pyruvate decarboxylase from brewer's yeast (*Saccharomyces carlsbergensis*) and commercial wheat germ (*Triticum aestivum*). The amino acid analyses were performed on nmol aliquots of 24, 48, and 72 hr hydrolysates with a Biotronik LC 6000 E amino acid analyzer using ninhydrin colorimetry and later *o*-phthalaldehyde fluorimetry for detection. The tryptophan values were obtained by photometry at 280/288 nm of the unresolved protein, denatured with 6 *M* guanidine.HCl. Cyst(e)ine was determined after oxidation with performic acid to cysteine sulfonic acid in separate runs.

Carboxypeptidase Y treatment revealed different C-terminal sequences for both chains of wheat germ PDC: α: . . . -Val-Ser-Ala-Leu; β: . . . -His-Asp-Ala-Ser. These results were still in favor of the hypothesis.

Detailed amino acid analyses, however, of either chain of PDC from both sources (Figure 1) turned out to be incompatible with the theory: for several amino acids, the differences between α and β chains were found greater than permitted by the mere removal of a short piece of chain of approximately 20 amino acids. In several cases, the α and β chains from either source differed more from each other than the corresponding chain from both sources. The most striking example is phenylalanine with an α to β ratio of 4:1 for both PDCs. In a few instances, the β chain had even a considerably higher content of the respective amino acid than the α chain: Ser and Gly for PDC from both sources, and Cys, Ala, and Arg for wheat germ PDC.

For further elucidation of the relatedness of α and β chains, an experimentally simple but conclusive peptide mapping method[24] was applied, as yet to only wheat germ PDC chains: proteolytic cleavage by *Streptococcus aureus* V8 protease on top of an electrophoretic gel gradient for a limited time, followed by electrophoretic separation and staining of the resulting peptides. Several peptides from each chain were found to be identical in size, but a comparable number of them exhibited quite different lengths.

The results of our experiments show that LM-PDC, the major fraction isolated from wheat germ, has a high similarity to yeast PDC in every respect investigated, that the α and β chains of PDC from both sources must be coded by different genes, but that these genes may have derived from one another during the evolution, and that the two genes are likely to be located on allelic chromosomes in brewer's yeast which is diploidic and in wheat (*Triticum aestivum*) which is a hexaploidic culture plant. Yeast PDC appears to be a little more adapted to its task in the organism than wheat germ PDC and PDC from other plants.

ACKNOWLEDGMENT

This work received financial support by grant No. Ul 21/9-2 from the Deutsche Forschungsgemeinschaft, D-5300 Bonn-Bad Godesberg.

REFERENCES

1. **Ullrich, J.**, *Meth. Enzymol.*, 18A, 109, 1970.
2. **Gounaris, A. D., Turkenkopf, I., Buckwald, S., and Young, A.**, *J. Biol. Chem.*, 246, 1302, 1971.
3. **Ludewig, R. and Schellenberger, A.**, *FEBS Lett.*, 45, 340, 1974.
4. **Ullrich, J. and Freisler, H.**, *Hoppe-Seyler's Z. Physiol. Chem.*, 358, 318, 1977.
5. **Holzer, H.**, *Adv. Enzyme Regul.*, 13, 125, 1975.
6. **Ullrich, J.**, *Ann. N.Y. Acad. Sci.*, 378, 287, 1982.
7. **Singer, T. P. and Pensky, J.**, *J. Biol. Chem.*, 196, 375, 1952.
8. **Laemmli, U. K.**, *Nature (London)*, 277, 680, 1970.
9. **Görg, A., Postel, W., Westermeier, R., Righetti, P. G., and Ek, K.**, LKB Application Note 320, LKB Produkter AB, S-16126 Bromma, 1981, 9.
10. **Nimmo, H. G. and Nimmo, G. A.**, *Anal. Biochem.*, 121, 17, 1982.
11. **Zehender, H., Trescher, D., and Ullrich, J.**, *Anal. Biochem.*, 135, 16, 1983.
12. **Feigl, F.**, *Spot Tests in Organic Analysis*, 5th ed., Elsevier, Amsterdam, 1956, 208.
13. **Kenworthy, P. and Davies, D. D.**, *Phytochemistry*, 15, 279, 1976.
14. **Gounaris, A. D., Turkenkopf, I., Civerchia, L. L., and Greenlie, J.**, *Biochim. Biophys. Acta*, 405, 492, 1975.
15. **Hopmann, R. F. W.**, *Eur. J. Biochem.*, 110, 311, 1980.
16. **Sieber, M., König, S., Hübner, G., and Schellenberger, A.**, *Biomed. Biochim. Acta*, 42, 343, 1983.
17. **Hübner, G., Weidhase, R., and Schellenberger, A.**, *Eur. J. Biochem.*, 92, 175, 1978.
18. **Righetti, P. G. and Gianazza, E.**, in *Electrophoresis*, Radola, B. J., Ed., Walter de Gruyter, Berlin, 1979, 23.
19. **Ullrich, J. and Donner, I.**, *Hoppe-Seyler's Z. Physiol. Chem.*, 351, 1030, 1970.
20. **Ostrovskii, Y., Ullrich, J., and Holzer, H.**, *Biokhimiya*, 36, 739, 1971; Engl. transl., p. 621, 1971.
21. **Ullrich, J., Ortlieb, E., and Weidner-Kleine, B.**, 12th FEBS Meeting, Abstr. No. 2865, Dresden, 1978.
22. **Schmitt, H. D. and Zimmermann, F. K.**, *J. Bacteriol.*, 151, 1146, 1982.
23. **Schmitt, H. D. and Zimmermann, F. K.**, *J. Mol. Gen. Genet.*, 192, 247, 1983.
24. **Cleveland, D. W., Fisher, S. G., Kirschner, M. W., and Laemmli, U. K.**, *J. Biol. Chem.*, 252, 1102, 1977.

Chapter 9

NOVEL FINDINGS ON BREWER'S YEAST PYRUVATE DECARBOXYLASE

Frank Jordan, Olufemi Akinyosoye, George Dikdan, Zbigniew Kudzin, and Donald Kuo

TABLE OF CONTENTS

I. INTRODUCTION

Although the enzyme yeast pyruvate decarboxylase (PDC, EC 4.1.1.1) has been known for a long time,[1] several aspects of its structure and mechanism still remain to be elucidated. The relatively slow progress can be partly attributed to the large size of the holoenzyme (about 250,000 daltons), and to the indiscriminate proteolytic activity characteristic of yeast, an activity that greatly diminishes long-term stability of the enzyme, and an activity that is only partially inhibited by the usual array of protease inhibitors. In this summary, we address some structural questions, such as the binding of cofactors and a variety of their analogs, the subunit compositions found for the brewer's yeast enzyme (two isozymes: one with a single, the other with two types of subunits), as well as cross-linking results on one of the isozymes. In addition, we address the question of the identification of an enzyme-bound enamine intermediate derived from a slow substrate *cum* inhibitor conjugated 2-keto acid, and the application of this observation to solve other mechanistic questions.

Scheme I presents the mechanism of PDC as understood on the basis of the seminal bioorganic model studies performed on thiamin derivatives by Breslow.[2]

Of the three key intermediates depicted on the reaction pathway, the 2-(1-hydroxyethyl)TDP, the 2-α-lactyl TDP, and the enamine (or 2-α-carbanion), the former two have been synthesized.[3,4] The third compound has never been detected either on or off the enzyme (i.e., produced by synthesis). Several years ago, we decided to pursue identification of such an enzyme-bound enamine structure. We believe that such a goal is well worth the effort since the enamine species is very likely present in all thiamin-diphosphate-dependent enzymatic reactions: it can be protonated at carbon to yield acetaldehyde (PDC); oxidized at the α-carbon by lipoamide (as in pyruvate dehydrogenases and other 2-keto acid dehydrogenases) or by FAD to yield acetate (as in pyruvate oxidase); or it can act as a nucleophile at the 2-α carbon to yield condensation products (as in transketolases). Earlier, we reported that the conjugated substrate analog (*E*)-4-(4-chlorophenyl)-2-keto-3-butenoic acid (*p*-CPB) when added to PDC not only inactivated the enzyme in a time-dependent fashion,[5] but also produced a new chromophore with an absorption maximum at 440 nm,[6] at very much longer wavelength than observed for the starting material, the expected product of turnover *p*-Cl-cinnamaldehyde), or a covalent adduct between the protein and either the conjugated starting material or conjugated product (such an adduct formed, e.g., by a Michael addition should interrupt the conjugated system, i.e., shift the absorption maximum to shorter rather than longer wavelengths). Based on its spectral and kinetic properties, we assigned the enamine structure derived from *p*-CPB to the chromophore.[6] Such an assignment, while difficult to make unequivocally, is also consistent with the spectroscopic properties reported in the literature on model compounds derived from C-2-substituted thiazolium derivatives,[7] and the fact that the 4′-amino group of TDP, the other potentially nucleophilic center of TDP toward the 2-keto group of *p*-CPB, has no detectable reactivity in a model system consisting of either *p*-CPB or *p*-Cl-cinnamaldehyde. 4-Amino-5-methoxymethyl-2-methylpyrimidine (a model for the aminopyrimidine portion of TDP) refluxed with BF$_3$-ether for 48 hr at 70°C. Below we present the results of some experiments that employed this spectroscopic observation in an attempt to (1) generate a PDC-based model for the first two reactions catalyzed by the 2-keto acid dehydrogenase multienzyme complexes, and (2) assign a role to a key cysteine of PDC.

II. MATERIALS AND METHODS

PDC was purified by a combination of methods[5] which included in turn a heat treatment, (NH$_4$)$_2$SO$_4$ fractionation, a second heat treatment, a treatment with alumina C-γ, and finally an ion-exchange chromatographic method. It was found that depending on the material

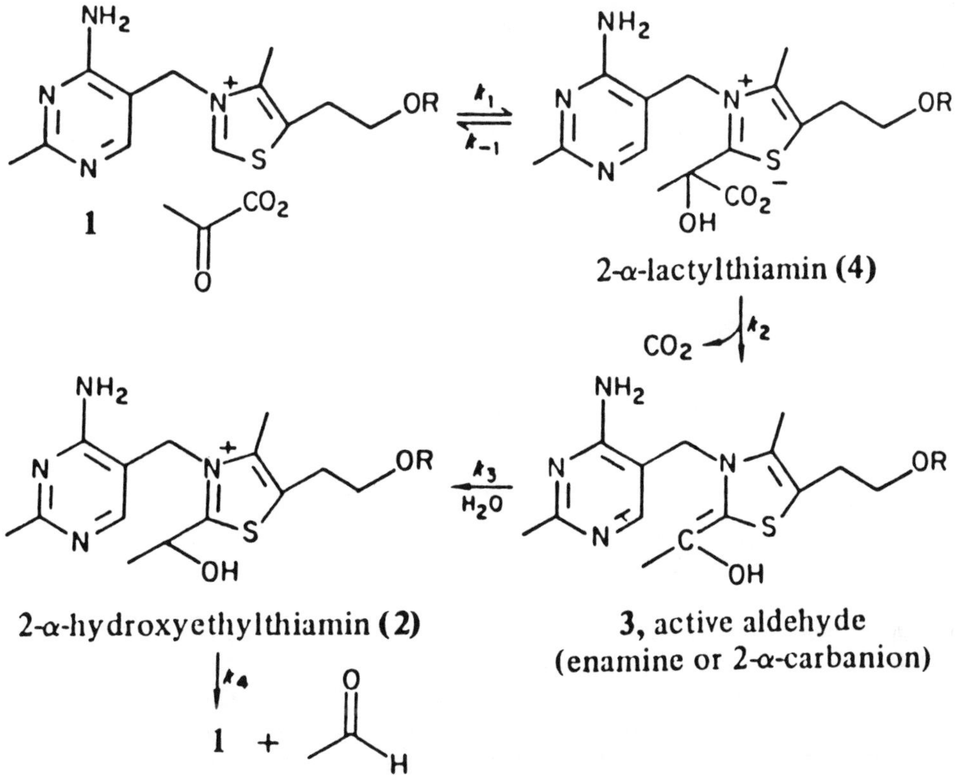

1

2-α-lactylthiamin (4)

CO_2 — k_2

2-α-hydroxyethylthiamin (2)

**3, active aldehyde
(enamine or 2-α-carbanion)**

k_3 / H_2O

k_1 / k_{-1}

k_4

1 +

SCHEME I. Mechanism of pyruvate decarboxylation catalyzed by thiamin diphosphate.

employed in the last step, one could emphasize one or the other active fraction containing PDC. The most efficient method employed DEAE-Sephadex for the last step and gave two fractions of PDC activity that were more than 85% pure according to polyacrylamide gel electrophoresis (performed according to Laemmli[8]).

The enzyme was assayed by the pH-stat method[9,10] and protein content was determined by Bradford's method.[11] One unit of activity is defined as the amount of enzyme required to convert 1 μmol of substrate to product per minute at 25°C, pH 6.0.

The conjugated 2-keto acid analogs were synthesized by condensing the appropriately substituted benzaldehyde with pyruvic acid under basic conditions.[5] The holoenzyme was resolved to the apo protein and the TDP and Mg^{2+} cofactors followed the procedure of Gounaris et al.:[12]

1. All inorganic chemicals were of the highest purity; water was double distilled and deionized.
2. Spectrofluorometric titrations were performed on either a Farrand MK1 or Spex 1902 Fluorolog spectrofluorometer.
3. Cross-linking was performed with either dimethylsuberimidate at pH 7.2[13] or glutar-aldehyde/$NaBH_3CN$ at pH 7.0.

III. RESULTS AND DISCUSSION

A. Subunit Structure and Cross-Linking

When purified using a DEAE-Sephadex ion-exchange column for the last step in the purification protocol, two readily distinguishable peaks elute with very sizable activity; in

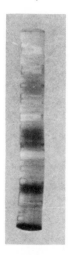

FIGURE 1.   Sodium dodecyl sulfate-polyacrylamide disc gel electrophoresis of cross-linked (glutaraldehyde, NaBH$_3$CN) yeast pyruvate decarboxylase. The molecular weights in ascending order are 54,000, 115,000, and 240,000 daltons.

order of elution from the column, these shall be referred to as fraction I and fraction II. By contrast, e.g., when DE-23 (Whatman) was employed, a much lower yield of purified material resulted which emphasized only fraction I as an electrophoretically pure species. Employing DEAE-Sephadex, the two fractions isolated are both more than 85 to 90% homogeneous according to disc gel electrophoresis. Fraction I consists of one band on SDS-PAGE (M$_r$ = 59,000 daltons), while fraction II consists of two bands (M$_r$ = 59,000 and 61,000 daltons). An approximately equimolar mixture of fractions I and II gives rise to only two bands, implying that the lower M$_r$ band of fraction II is similar if not identical to the protein in fraction I. Rechromatographing each fraction according to the same protocol produced no change in the appearance of the SDS-PAGE results on each fraction; i.e., the two fractions are not interconverted proteolytically at this stage. The two active fractions appear to be essentially the same so far as specific activities are concerned (47.5 and 41.8 U/mg, respectively). Both fractions exhibit a distinctly sigmoidal v$_o$ vs. [pyruvate] behavior providing K from the appropriate Hill plots of 0.85 mM (n = 2.33) and 0.95 mM (n = 2.11) for fractions I and II, respectively. In addition, both fractions also behave very similarly vis-à-vis the mechanism-based inactivator p-CPB.[5]

While we do not yet know whether or not the lower M$_r$ band is a proteolytic breakdown product of the higher M$_r$ band, when searching for a consistent interpretation, one must account for both the activity of fraction I (composed exclusively of lower M$_r$ subunits) and the fact that we have never observed activity from a species consisting exclusively of the higher M$_r$ subunits.

These observations resolve the controversy concerning the existence of identical[11,14] or nonidentical[15,16] subunits in PDC, but not the relationship, if any, between them.

In a further study, we attempted to cross-link PDC. For these studies, we employed PDC fraction I, i.e., the one consisting of identical subunits and purified by chromatography on DE-23 (Whatman). When apo PDC was allowed to react with dimethylsuberimidate at pH 7.2,[13] no cross-linked products could be detected by SDS-PAGE in either the absence or presence of the cofactors (as much as 1 mM TDP and Mg^{2+}). When the cross-linking was performed with glutaraldehyde at pH 7.1 followed by reduction with 2 mM NaBH$_3$CN, cross-linking resulted. On SDS-PAGE we observed species with M$_r$ of 55,000, 115,000, and 240,000 daltons (Figure 1), corresponding to monomers, dimers, and tetramers. Precisely the same patterns resulted irrespective of the absence or presence (as much as 1 mM TDP and Mg^{2+}) of the cofactors. From these results, we conclude that fraction I of PDC undergoes spontaneous association to dimers and tetramers (presumably composed of a dimer of dimers), and that this subunit association requires neither the substrate, nor the cofactors.

B. Association of Apo PDC with Cofactors and Cofactor Analogs

In order to assess the possibility of substituting cofactor analogs that may act as reporter groups of the active center of the enzyme, a series of substitutions were affected. The methodology employed involved first resolving the holoenzyme into apo protein and cofactors by adjusting the pH of the PDC solution to 8.0 in the presence of 1 mM EDTA and incubating at 4°C for 45 min,[12] then chromatographing the mixture on Sephadex G-25. HPLC, employing fluorescence detection of thiochromes (produced by CNBr oxidation), showed that TDP had been quantitatively removed. Reconstitution was performed by a stepwise change in the buffer from 8.0 to 7.5 to 7.0 to 6.5 and finally to 6.0, incubating for 0.5 hr in each solution (each of which also contained 2 mM dithiothreitol and 0.04 M KCl), then adding the cofactor or cofactor analog.

Thiamin thiothiazolone diphosphate (TTTDP) was reported to be a transition state analog for the *Escherichia coli* pyruvate dehydrogenase multienzyme complex.[18] We therefore decided to determine the K_d of this compound compared to that of TDP for PDC. A fluorescence titration of apo PDC with TDP or TTTDP indicated that very likely there are two classes of binding constants for both: one in the micromolar, the other in the millimolar range (fluorescence excitation at 280 nm; emission at 343 nm). The stronger K_d had a value of 6×10^{-6} M for TDP and virtually the same value (7×10^{-6} M) for TTTDP. Prior saturation of the enzyme with Mg^{2+} did not affect the TDP binding curve at all. Based on these results, we conclude that TTTDP is not a transition state analog of PDC, just as its oxygen analog is not a tight-binding analog for pyruvate oxidase.[19] However, it was found that TTTDP red-shifted the emission maximum of the apo PDC by 12 nm. In a relevant model study we had observed that on binding thiamin thiothiazolone to tryptophan, there is a red-shift in the emission maximum of the tryptophan by 7 nm.[20] The observation on the binding of TTTDP to PDC apparently involves binding to a tryptophan on the enzyme. The red-shift observed on binding hydrophobic molecules to tryptophan has other precedents.[21] Although there have been model studies in favor of such binding,[20,22-24] to our best knowledge, the above observation is probably the most direct demonstration of such an interaction between a cofactor analog and the enzyme.

Next, we attempted to replace the essential Mg^{2+} with lanthanide ions with the expectation that these would behave as reporter groups. Reconstitution of the apo PDC with Eu^{3+} or Tb^{3+} in place of Mg^{2+} afforded approximately 30% of the activity of the unresolved holo PDC compared to the 50% activity achieved with Mg^{2+}. Two qualitative results are noteworthy. Upon adding incremental concentrations of Eu^{3+} to apo PDC there was a dramatic, nearly 10,000-fold, increase in the fluorescence intensity attributed to the Eu^{3+} (excitation at 395 nm; emission measured at 580 nm). Similarly impressive observations were made on adding Tb^{3+} to the apo PDC (excitation at 280 nm; emission measured at 480 nm). These enormous fluorescence enhancements enabled us to conclude that all these observations pertained to the apo PDC-lanthanide ion binary complexes. It was also observed that the fluorescence emission maxima of the lanthanide ions suffered a very sizable blue-shift on the protein surface when compared to the emission maxima measured under the same conditions but in water (15 nm for Eu^{3+} and 10 nm for Tb^{3+}). These fluorescence enhancements are due to transfer of energy from a protein tryptophan to the metal ion.[25] The blue-shifts in emission maxima are consistent with transfer of the lanthanide ion from an aqueous to a hydrophobic protein environment.[25] While there have been previous reports of a hydrophobic active center on PDC based on fluorescence measurements on the binding of the competitive inhibitor ANS,[26,27] this report provides the first evidence for a hydrophobic environment around the metal binding locus.

C. Experiments with the Conjugated 2-Keto Acid Mechanism-Based Inhibitors

We had reported earlier that *p*-CPB inactivates PDC in a time-dependent fashion, and the kinetics of inactivation imply two-site interaction presumably at the regulatory and catalytic sites, prior to inactivation at the catalytic site.[5] When we employed [1-^{14}C]-*p*-CPB, radioactive $^{14}CO_2$ was released. By contrast, when we employed [3-^{3}H]-*p*-CPB,[38] the tritiated material remained with the protein even upon extended dialysis, even after incubation above pH 8.0, a condition under which the cofactors are released, and even upon heating in SDS for the PAGE experiments. These results suggest that the enamine structure derived from the decarboxylation of the conjugated 2-keto acid on PDC-bound TDP forms a covalent bond with the protein, perhaps by the reaction of a nucleophilic side chain of the protein with the enamine or some chemically rearranged product therefrom. The site of labeling is not yet determined, nor is the stoichiometry of inhibitor to cofactor to protein subunit. In addition, evidence for the formation of the enamine was also obtained from VIS spectroscopic experiments. Addition of *p*-CPB to PDC·TDP produced a new chromophore at 440 nm that could be attributed to the enamine structure (see Reference 6 and the evidence presented in Section I) depicted in the structure:

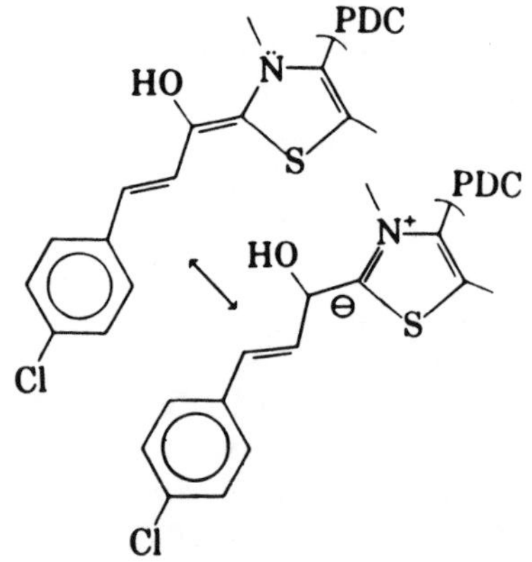

Assignment of this absorbance to the enamine allowed us to answer some mechanistic questions regarding both PDC and the related pyruvate dehydrogenase multienzyme complex.

D. PDC-Generated Model for the Reductive Acyl Transfer Catalyzed by 2-Keto Acid Dehydrogenases

The 2-keto acid dehydrogenase multienzyme complexes catalyze a series of enzymatic reactions whose most important products are acylCoA compounds.[28] The first two steps of the sequence as usually written for pyruvate are as follows (and acknowledging the uncertainty surrounding the true nature of the key intermediate):

$$E1.TDP + CH_3COCO_2H \rightarrow E1.2\text{-}(1\text{-hydroxyethyl})\text{-}TDP$$
$$\text{or } E1\text{-``enamine'' (from Scheme I)} \qquad (1)$$

$$E1.2\text{-}(1\text{-hydroxyethyl})\text{-}TDP(\text{or } E1\text{-``enamine''}) + E2\text{-Lipoamide} \rightarrow \qquad (2)$$
$$E1.TDP + E2.acetyldihydrolipoamide$$

The observation of the enamine derived from *p*-CPB enabled us to test the mechanism of the reaction sequence described in Equations 1 and 2 in an enzyme-generated model system.

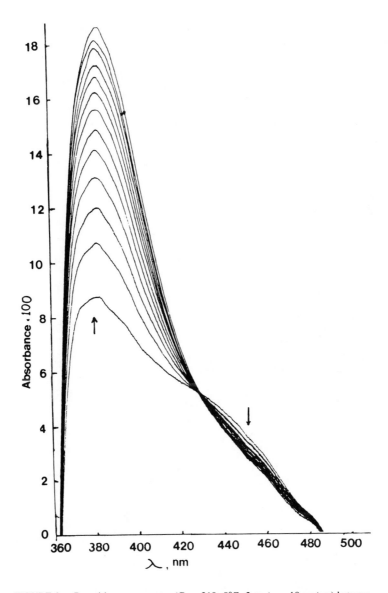

FIGURE 2. Repetitive scan spectra (Cary 219, 8°C, 2 nm/sec, 10 nm/cm) between 360 and 500 nm of the trapping of the PDC-bound enamine (resulting from decarboxylation of *p*-CPB) by 4,4′-dithiodipyridine. The scanning was performed for 23 min after the addition of 100 μℓ of 10 mM 4,4′-dithiodipyridine to a solution of 700 μℓ "mix" that contained 0.1 M pyrophosphate, 1 mM TDP, 1 mM MgSO$_4$, 0.5 mM EDTA, 0.5 mM phenylmethanesulfonyl fluoride, 20% ethylene glycol (v/v) in 0.1 M citrate, pH 6.0, along with 40 U PDC and 100 μℓ of 20 mM *p*-CPB which was previously incubated for 50 min at 8°C. The reference cuvette contained 750 μℓ of the "mix" and 50 μℓ of 20 mM CPB. The arrows indicate the direction of the change in absorbance with time.

We would first generate the enamine from *p*-CPB according to the development of A$_{440}$ (mimicking Equation 1), then trap the enamine with a disulfide, in a reaction analogous to Equation 2. For this purpose, 4,4′-dithiodipyridine (DTDP) was selected in place of lipoic acid. First of all, linear disulfides have a greater tendency to be reduced; second, DTDP is highly chromophoric in its reduced state and, we expected, so would any covalent intermediates that would be observed during the reaction. Figure 2 shows the time course of

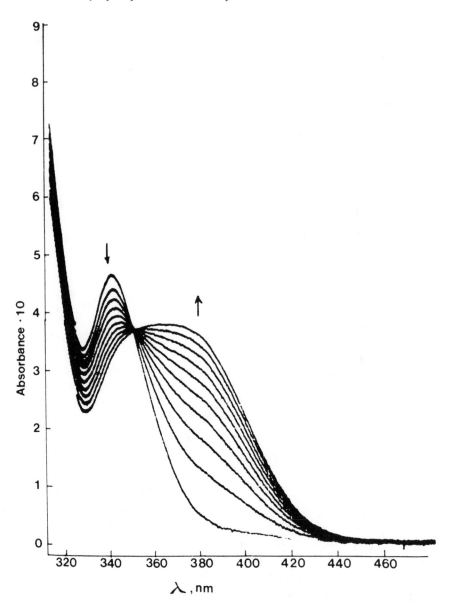

FIGURE 3.   Repetitive scan spectra (Cary 219, 4°C, 2 nm/sec, 10 nm/cm) between 310 and 480 nm of the trapping of the PDC-bound enamine (from pyruvic acid) by 4,4′-dithiodipyridine. The reaction was initiated by adding 50 μℓ of 0.5 M pyruvic acid to a solution containing 600 μℓ of the "mix" in Figure 2, 200 μℓ (25 U) PDC, and 100 μℓ of 10 mM 4,4′-dithiopyridine. The scanning was continued for 19 min after the initiation of the reaction. The reference cuvette contained 800 μℓ of the "mix" in Figure 2 and 50 μℓ of 10 mM p-CPB.

development of the VIS spectrum resulting from the mixing of DTDP with enamine derived from p-CPB: the absorbance corresponding to the enamine at 440 nm diminishes with time, but a new absorbance with maximal absorbance near 380 nm develops. While the A_{440} could be demonstrated to pertain to an enzyme-bound species that is formed in stoichiometric amounts, the A_{380} was formed continuously, i.e., accumulated off the enzyme. It was next desirable to determine if the A_{380} could be produced from only the conjugated 2-keto acid p-CPB, or indeed from any 2-keto acids. Figure 3 presents the results of an experiment in which pyruvate, PDC·TDP, and DTDP were all mixed together at the initiation of the

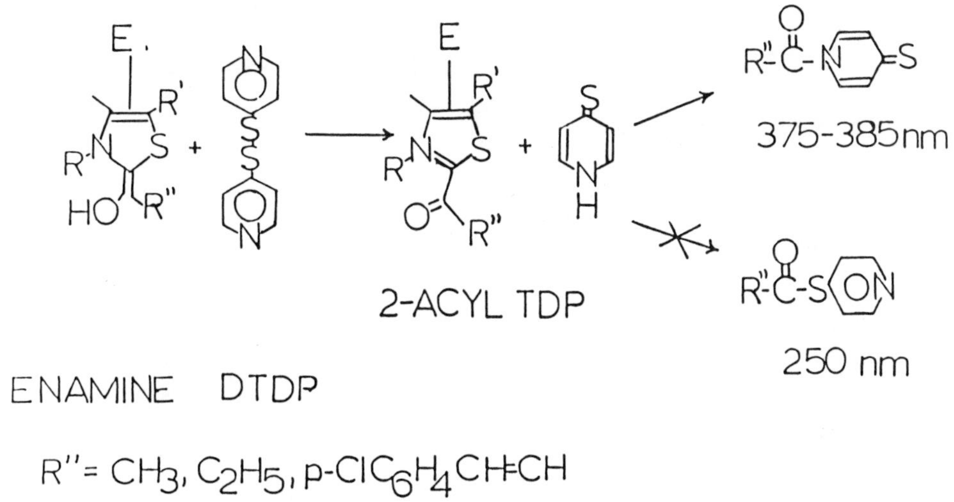

2-ACYL TDP

ENAMINE   DTDP

$R'' = CH_3, C_2H_5, p\text{-}ClC_6H_4 CH\!\!=\!\!CH$

SCHEME II. Mechanism for a stepwise oxidative acyl transfer from the enamine to the reduced thiol produced by 4,4'-dithiodipyridine.

observation. Very clearly, a new absorbance centered near 380 nm again resulted (in this case, the absorbance due to the enamine is not observable for a variety of reasons). By rigorous structure determination, it was proved that the absorbances produced by following such a protocol for a variety of 2-keto acids (including 2-keto butyric) all had maxima between 370 to 385 nm, and pertained to the N-acylated 4-thiopyridone structure:

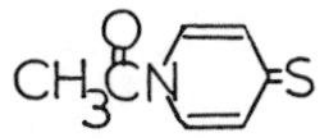

Apparently, the acyl group had only minor effect on the absorbance maxima observed, and the chromophore reflected the electronic properties of the 4-thiopyridone. In order to interpret these results, one more experiment is relevant: under a variety of conditions examined (including nonaqueous medium), the model compound 2-(1-hydroxyethyl)-3,4-dimethyl-thiazolium ion was *totally resistant* to oxidation by DTDP. This suggested that the true intermediate for oxidation in both the model thiamin reactions and our PDC-generated model system is the enamine structure rather than the 2-(1-hydroxyethyl) derivative in Equations 1 and 2. A mechanism consistent with the results is presented in Scheme II.

The enamine intermediate is oxidized (without prior protonation) to the 2-acyl TDP derivative (the reaction produces 4-thiopyridone as the reduction product), which can be deacylated by either the N or S end of the tautomeric nucleophile 4-thiopyridone. Apparently, the N deacylates 2-acyl TDP preferentially, hence the observed product. This model system provides several pieces of useful data: (1) the spectral behavior observed in conjunction with the chemistry taking place gives strong supporting evidence for the assignment of A_{440} to the enamine intermediate;[6] (2) the identification of the product of the oxidation reaction performed by DTDP on the enamine gives very good evidence for the intermediacy of 2-acyl TDP on the pathway in this PDC-generated model for Equations 1 and 2, and suggests at least the possibility for a similar stepwise reductive acylation of E2-lipoamide in the 2-keto dehydrogenase multienzyme complexes as well. It is relevant that a similar stepwise mechanism was also suggested recently for 2-ketoglutarate dehydrogenase.[29]

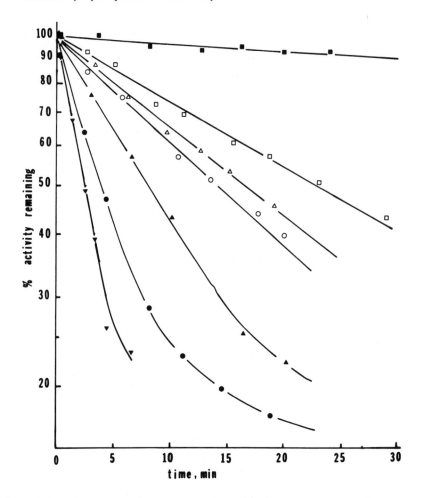

FIGURE 4.   Inactivation of brewer's yeast pyruvate decarboxylase by methyl methanethiol-sulfonate (MMTS). Samples (0.5 mℓ) containing 11 U of PDC, 1 mM TDP, and 2 mM MgSO$_4$ in 0.1 M maleate, pH 6.0, were incubated at 30°C with varying concentrations of MMTS: (■) 0 mM; (□) 0.2 mM; (△) 0.25 mM; (○) 0.3 mM; (▲) 0.5 mM; (●) 1.0 mM; (▼) 2 mM. At the indicated time intervals, 50-$\mu\ell$ aliquots were removed for assay by the pH-stat method.

E. Role of a Cysteine in the Catalytic Mechanism

Reports from a number of laboratories have indicated that at least one Cys–SH group has vital importance in the mechanism of action of PDC[30-34] (the holoenzyme possesses as many as 20 cysteines per 250,000 daltons[34]). The spectroscopic tool provided by the observation of the enamine derived from p-CPB enabled us to address the step in the mechanism (as depicted in Scheme I) which would be susceptible to –SH modification. The working hypothesis was that if PDC that had been inactivated by –SH modification could still produce enamine from p-CPB, the Cys being modified would have a role in the conversion of the enamine to product. If the –SH-modified inactive enzyme could no longer produce the A$_{440}$ characteristic of the enamine derived from p-CPB,[6] the Cys being modified would have a role in the formation of the enamine. As a sulfhydryl reagent, methylmethanethiol sulfonate (MMTS) offers some advantages,[35] such as delivery of the small –SCH$_3$ group onto an –SH, a modification that would introduce perhaps only a minor steric perturbation in the active center.

Figure 4 demonstrates that methyl methanethiolsulfonate (MMTS) inactivates PDC in a

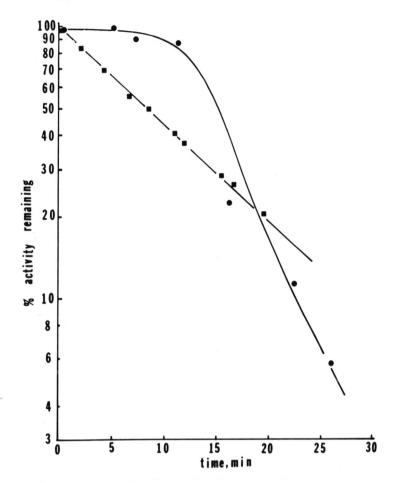

FIGURE 5. The time course of inactivation of PDC by MMTS in the absence and presence of pyruvate. Samples (0.5 mℓ) contained 0.8 mM MMTS, 1 mM TDP, 20 mM K$_4$P$_2$O$_7$, and 1 mM MgSO$_4$ in 0.1 M citrate, pH 6.0, 30°C in the absence (■) and presence (●) of 60 mM pyruvate. At the indicated time intervals, 50-μℓ aliquots were removed and assayed for PDC activity by the pH-stat method.

time-dependent fashion. The double-reciprocal plot[36] of T$_{1/2}$ for inactivation (obtained from the initial linear portions of the curves in Figure 4) vs. [MMTS]$^{-1}$ was linear, and extrapolated to the origin at infinite MMTS concentration, indicating that the interaction between PDC and MMTS followed second-order kinetics and did not involve saturation. Pyruvamide (at 40 mM concentration), a known allosteric effector of PDC,[37] had no effect on the inactivation by MMTS. Pyruvate (60 mM), on the other hand (Figure 5), initially afforded complete protection from MMTS-induced inactivation. Once the pyruvate concentration diminished significantly due to turnover, the inactivation proceeded very effectively. We interpret these results to mean that MMTS inactivates the enzyme at its catalytic center (this by no means implies that all sulfhydryl reagents will affect PDC in an analogous fashion).

When PDC was first inactivated with MMTS for 30 min at 2.2°C, then the p-CPB was added; the magnitude of the increase in A$_{440}$ was identical to that observed in the absence of prior MMTS-induced inactivation (Figure 6). This result suggests that the −SH group here modified by MMTS has a role in a step past decarboxylation, i.e., in the conversion of the enamine to product. The particular −SH here modified (at the catalytic center) cannot have the role previously suggested,[32] i.e., the binding of the 2-keto acid as a hemiketal. It is also noteworthy that when p-CPB was added to PDC that had first been inactivated with

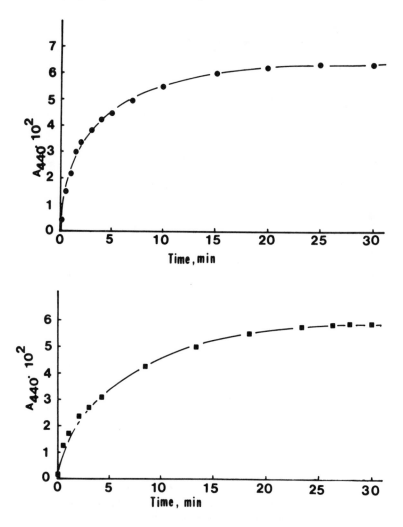

FIGURE 6.   The interaction of *p*-CPB with PDC with and without prior inactivation by MMTS. To 31 U of PDC in 580 μℓ of the "mix" (of Figure 2) was added 150 μℓ of 10 m*M* MMTS and the mixture was incubated for 30 min at 2.2°C. Next, 70 μℓ of 30 m*M* *p*-CPB was added, and the increase in A_{440} was recorded with time (●). The control experiment (■) omitted the MMTS treatment (and incubation period), but employed 150 μℓ water instead.

MMTS (conditions under which the enamine is still formed, as indicated by the build-up of A_{440}), DTDP could no longer oxidize the enamine: the A_{440} persisted; the A_{380} did not develop.

Therefore, the methylthiolated −Cys either hinders DTDP from approaching the enamine, or the −SH being modified by MMTS is an integral participant in the oxidation of the enamine performed by DTDP. In any case, the weight of the evidence supports the contention that the Cys modified by MMTS is located in the catalytic center. Its function could be that of a general acid or general base catalyst, required in the protonation of the enamine or in the deprotonation of the hydroxy group to promote expulsion of acetaldehyde from the 2-(1-hydroxyethyl)TDP, respectively.

F. Conclusions and Prospects for Future Research

The results presented above convinced us to pursue two lines of inquiry, both of which are still in preliminary stages.

1. The observation of two "isozyme"-like fractions during the purification by DEAE-Sephadex chromatography requires further study. This will be much aided by the rumored recently completed genetic structure of brewer's yeast PDC.
2. The observation of the enamine intermediate and the experiments that were completed employing that observation will be expanded in several directions. For example, we have undertaken a synthetic program to generate structures related to the enamine in Scheme I for spectroscopic structure proof. Also, we have made a number of analogs of *p*-CPB which bear different substituents on the phenyl ring. Extensive studies indicated that the *m*-NO$_2$ analog binds considerably better than *p*-CPB to PDC. The time course of formation and disappearance of the enamine derived from the *m*-NO$_2$ analog could be monitored by VIS spectroscopy (maximum absorbance centered at 430 nm). The kinetic resolution of this complicated behavior will provide details about some of the elementary steps shown in Scheme I. Finally, extension of the methodology to aliphatic 2-keto acids (employing rapid kinetic techniques) will provide information about the mechanistic fate of more substrate-like compounds.

ACKNOWLEDGMENTS

We are grateful to Professor Alfred Schellenberger for organizing the Second Thiamin Symposium and for being a most gracious host at Wernigerode, GDR, in October 1984.

We express our thanks for financial support of our research on thiamin chemistry and enzymology to NIH-AM-17495, NSF-PCM 8217100, the Charles and Johanna Busch Fund at Rutgers University, and Hoffmann-LaRoche, Inc., Nutley, N.J.

Lastly, we also express thanks to the Anheuser Busch Brewing Co., Newark, N.J., for their continued generosity in providing us with the brewer's yeast.

REFERENCES

1. **Krampitz, L. O.**, *Thiamin Diphosphate and Its Catalytic Functions*, Marcel Dekker, New York, 1970, chap. 2.
2. **Breslow, R.**, On the mechanism of thiamine action. IV. Evidence from studies on model systems, *J. Am. Chem. Soc.*, 80, 3719, 1958.
3. **Krampitz, L. O. and Votaw, R.**, 2-(1-Hydroxyethyl)thiamine diphosphate and 2-(1,2-dihydroxyethyl)thiamine diphosphate, *Methods Enzymol.*, 9, 65, 1966.
4. **Kluger, R. and Smyth, T.**, Interaction of pyruvate-thiamin diphosphate adducts with pyruvate decarboxylase. Catalysis through "closed" transition states, *J. Am. Chem. Soc.*, 103, 1214, 1981.
5. **Kuo, D. J. and Jordan, F.**, Active-site-directed irreversible inactivation of brewer's yeast pyruvate decarboxylase by the conjugated substrate analogue (E)-4-(4-chlorophenyl)-2-oxo-3-butenoic acid: development of a suicide substrate, *Biochemistry*, 22, 3735, 1983.
6. **Kuo, D. J. and Jordan, F.**, Direct spectroscopic observation of a brewer's yeast pyruvate decarboxylase-bound enamine intermediate produced from a suicide substrate, *J. Biol. Chem.*, 258, 13415, 1983.
7. **Larivé, H. and Dennilauler, R.**, Cyanine dyes derived from thiazolium salts, in *The Chemistry of Heterocyclic Compounds*, Vol. 34 (Part 3), Metzger, J., Ed., John Wiley & Sons, New York, 1979, chap. 9.
8. **Laemmli, U. K.**, Cleavage of structural proteins during the assembly of the head of bacteriophage T4, *Nature (London)*, 227, 680, 1970.
9. **Leussing, D. L. and Stanfield, C. K.**, Kinetics and formation of N-pyruvylidenyglycinatozinc (II), *J. Am. Chem. Soc.*, 88, 5726, 1966.
10. **Schellenberger, A., Hubner, G., and Lehmann, H.**, A new test for the determination of pyruvate decarboxylase activity, *Angew. Chem. Int. Ed. (Engl.)*, 7, 886, 1968.
11. **Bradford, M. M.**, A rapid and sensitive method for quantitation of microgram quantities of protein utilizing the principle of protein-dye binding, *Anal. Biochem.*, 72, 248, 1976.

12. **Gounaris, A. D., Turkenkopf, I., Buckwald, S., and Young, A.,** Pyruvate decarboxylase. I. Protein dissociation into subunits under conditions in which thiamine pyrophosphate is released, *J. Biol. Chem.,* 246, 1302, 1971.

13. **Davies, G. E. and Stark, G. R.,** Use of dimethyl suberimidate, a cross-linking reagent, in studying the subunit structure of oligomeric proteins, *Proc. Natl. Acad. Sci. U.S.A.,* 66, 651, 1970.

14. **Ludewig, R. and Schellenberger, A.,** A new procedure to prepare highly purified and crystallized yeast pyruvate decarboxylase, *FEBS Lett.,* 45, 340, 1974.

15. **Hopmann, R. F. W.,** Hydroxyl ion-induced subunit dissociation of yeast cytoplasmic pyruvate decarboxylase, *Eur. J. Biochem.,* 110, 311, 1980.

16. **Sieber, M., Koenig, S., Hübner, G., and Schellenberger, A.,** A rapid procedure for the preparation of highly purified pyruvate decarboxylase from brewer's yeast, *Biomed. Biochim. Acta,* 42, 343, 1983.

17. **Gounaris, A. D., Turkenkopf, I., Civerchia, L. L., and Greenlie, J.,** Pyruvate decarboxylase. III. Specificity restrictions for thiamin pyrophosphate in the protein association step, sub-unit structure, *Biochim. Biophys. Acta,* 405, 492, 1975.

18. **Gutowski, J. A. and Lienhard, G. E.,** Transition state analogs for thiamin pyrophosphate-dependent enzymes, *J. Biol. Chem.,* 251, 2863, 1976.

19. **O'Brien, T. A. and Gennis, R. B.,** Studies of the thiamin pyrophosphate binding site of *E. coli* pyruvate oxidase, *J. Biol. Chem.,* 255, 3302, 1980.

20. **Farzami, B., Mariam, Y. H., and Jordan, F.,** Solvent effects on thiamin-enzyme model interactions. I. Interactions with tryptophan, *Biochemistry,* 16, 1105, 1977.

21. **Burshtein, E. A., Vedenkina, N. S., and Ivkova, M. N.,** Fluorescence and the location of tryptophan residues in protein molecules, *Photochem. Photobiol.,* 18, 263, 1973.

22. **Sable, H. Z. and Biaglow, J. E.,** Coenzyme interactions: proton magnetic resonance study of molecular complexes of thiamine and indole derivatives, *Proc. Natl. Acad. Sci. U.S.A.,* 54, 808, 1965.

23. **Biaglow, J. E., Mieyal, J. J., Suchy, J., and Sable, H. Z.,** Coenzyme interactions. III. Characteristics of the molecular complexes of thiamine with indole derivatives, *J. Biol. Chem.,* 244, 4054, 1969.

24. **Mieyal, J. J., Suchy, J., Biaglow, J. E., and Sable, H. Z.,** Coenzyme interactions. IV. Molecular complexes of thiamine with indole derivatives — quantitative aspects, *J. Biol. Chem.,* 244, 4063, 1969.

25. **Horrocks, W. De W., Jr. and Sudnick, D. R.,** Lanthanide ion luminescence probes of the structure of biological macromolecules, *Acc. Chem. Res.,* 14, 384, 1981.

26. **Ullrich, J. and Donner, I.,** Fluorimetric study of 2-p-toluidino-naphthalene-6-sulfonate binding to cytoplasmic yeast pyruvate decarboxylase, *Hoppe-Seyler's Z. Physiol. Chem.,* 351, 1030, 1970.

27. **Ullrich, J., Ostrovskii, Yu. M., Eyzaguirre, J., and Holzer, H.,** Thiamine pyrophosphate-catalyzed enzymic decarboxylation of 2-oxo acids, *Vitam. Horm. N.Y.,* 28, 365, 1970.

28. **Reed, L. J.,** Multienzyme complexes, *Acc. Chem. Res.,* 7, 40, 1974.

29. **Steginsky, C. A. and Frey, P. A.,** *E. coli* 2-ketoglutarate dehydrogenase complex; thiamin diphosphate-dependent hydrolysis of succinyl coenzyme A, *J. Biol. Chem.,* 259, 4023, 1984.

30. **Stoppani, A. O. M., Actis, A. S., Deferrari, J. O., and Gonzalez, E. L.,** The role of sulfhydryl groups of yeast carboxylase, *Biochem. J.,* 54, 378, 1953.

31. **Kanopkaite, S. I.,** Active sulfhydryl group components of carboxylase, *Biokhimiya,* 21, 834, 1956.

32. **Schellenberger, A. and Hübner, G.,** Binding of the substrate in yeast pyruvate decarboxylase, *Angew. Chem. Int. Ed. (Engl.),* 7, 68, 1968.

33. **Brauner, T. and Ullrich, J.,** Yeast pyruvate decarboxylase. Number and reactivity of mercapto groups, *Hoppe-Seyler's Z. Physiol. Chem.,* 353, 825, 1972.

34. **Ullrich, J.,** Structure-function relationships in pyruvate decarboxylase of yeast and wheat germ, *Ann. N.Y. Acad. Sci.,* 378, 287, 1982.

35. **Smith, D. J., Maggio, E. T., and Kenyon, G. L.,** Simple alkanethiol groups for temporary blocking of sulfhydryl groups of enzymes, *Biochemistry,* 14, 766, 1975.

36. **Kitz, R. and Wilson, I. B.,** Esters of methanesulfonic acid as irreversible inhibitors of acetylcholinesterase, *J. Biol. Chem.,* 237, 3245, 1962.

37. **Hübner, G., Weidhase, R., and Schellenberger, A.,** The mechanism of substrate activation of pyruvate decarboxylase: a first approach, *Eur. J. Biochem.,* 92, 175, 1978.

38. **Jordan, F., Akinyosoye, O., Dikdan, G., Kudzin, Z., and Kuo, D.,** unpublished results.

Chapter 10

TRANSITION STATES IN DIFFERENT REACTIONS CATALYZED BY PDC

Gerhard Hübner

Besides the physiological reaction of the decarboxylation of 2-oxo acids (Scheme IA), pyruvate decarboxylase (EC 4.1.1.1) (PDC) catalyzes in the presence of H acceptors also oxidative decarboxylation with formation of the corresponding carbon acids (Scheme IB).

Another interesting reaction is observed if 3-halogen-2-oxo acids are metabolized by PDC. In this reaction, the substrate is converted to acetate while halogen elimination and decarboxylation take place (Scheme IC). In all three reactions, PDC displays only small substrate specificity. Therefore, an attempt was undertaken to obtain information on the mechanism of the catalysis reaction and the structure of the transition state by investigating the influence of substituent parameters of the substrate on the PDC-catalyzed reactions.

First, the influence of substituent parameters on the decarboxylation and the oxidative decarboxylation proceeding in the presence of 2,6-dichlorophenolindophenol (DCPIP) is considered.

The native reaction was observed with the help of the pH-stat test, whereas the oxidative decarboxylation was measured spectroscopically by bleaching of the H acceptor at 600 nm. Figure 1 exhibits the result of the investigations. The maximum rates for both the nonoxidative decarboxylation (open symbols) and the oxidative decarboxylation (filled symbols) show a good coincidence for this type of substrate. Furthermore, they yield an LFE-relationship to Hammett's substituent constants. This outcome indicates a common rate-limiting transition state for oxidative and nonoxidative decarboxylation until the formation of the α-carbanion. According to O'Leary,[1] the decarboxylation is not rate limiting. Therefore, for the substrate class of phenyl glyoxylic acids (PGA), a reaction step must be rate limiting which is connected with the formation of the carbonyl bond. The substituent effect, too, speaks in favor of such a rate-limiting step.

Corresponding with the observation that in carbonyl addition reactions the rate of addition of a nucleophilic group is enhanced by electron-withdrawing substituents, 4-substituted PGA with positive σ value are metabolized faster.

Another interesting result was attained with a PDC preparation containing the modified coenzyme 4'-hydroxy-thiamin pyrophosphate (TPP). The native reaction is known to display a blocked catalysis mechanism if the 4'-NH_2 group of the coenzyme is changed. As Table 1 shows, oxidative decarboxylation is also blocked if 4'-hydroxy-TPP is incorporated into PDC instead of native TPP. This result indicates that the 4'-amino group also has a function in a reaction step prior to the differentiation of the two reactions.

Concerning the reaction of the decarboxylation of 3-halogen pyruvate proceeding via halogen elimination, if chloropyruvate is incubated with PDC, a catalytic reaction can be observed by CO_2 formation in the manometric test as well as by pH-stat titration or chloride determination with an ion-specific electrode. Figure 2 demonstrates that the formation of CO_2, hydrogen ions, and chloride ions occurs in stochiometric ratios.

A product analysis shows that only acetate can be detected as further reaction product. Therefore, a reaction can be formulated, corresponding to a halogen-eliminating oxidative decarboxylation of the substrate. Analogously, fluoro-, bromo-, dichloro-, and dibromo-pyruvate are metabolized, whereas hydroxypyruvate is decarboxylated in the normal reaction, yielding glycolaldehyde.

The sigmoidal V/S plot obtained in the case of the normal PDC reaction with pyruvate as substrate as well as the typical lag phase during product formation — both being results

$$R-CO-COO^- + H_2O \xrightarrow{\text{PDC}} R-CHO + CO_2 + OH^- \qquad \text{A}$$

$$R-CO-COO^- + DCIP + H_2O \xrightarrow{\text{PDC}} R-COO^- + CO_2 + DCIP \cdot H_2 \qquad \text{B}$$

$$XCH_2-CO-COO^- + H_2O \xrightarrow{\text{PDC}} CH_3-COO^- + CO_2 + X^- + H^+ \qquad \text{C}$$

SCHEME I.

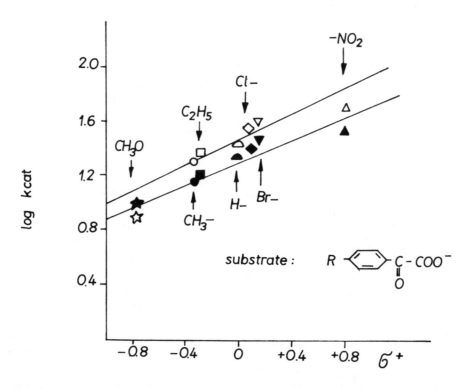

FIGURE 1.   Dependence of the catalytic constants for the PDC-catalyzed decarboxylation (open symbols) and the oxidative decarboxylation (filled symbols) of 4-substituted PGA on the substituent constants according to Brown and Okamoto.

Table 1

INFLUENCE OF THE COENZYME ON DIFFERENT REACTIONS CATALYZED BY PDC (0.1 *M* CITRATE BUFFER, pH 6.2 T = 25°C)

		k_{cat} (sec^{-1})		
Coenzyme	**Substrate**	**Decarboxylation**	**Oxidative decarboxylation**	**Halogen-eliminating decarboxylation**
TPP	Pyruvate	316	3.83	—
4'-Hydroxy-TPP	Pyruvate	0	0	—
TPP	Chloropyruvate	0	0.24	85
4'-Hydroxy-TPP	Chloropyruvate	0	—	0

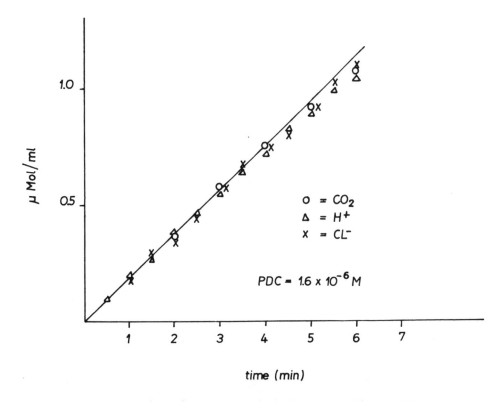

FIGURE 2. Progress curve of the conversion of chloropyruvic acid by PDC.

of the PDC activation by its substrate — can be observed with all halogen pyruvates as substrates (Figure 3). Table 2 summarizes the K_M and k_{cat} values obtained for this type of substrate.

Provided that the rate of conversion of halogen pyruvic acids is controlled by a step connected with the elimination of the halogen ion, the reaction rate should decline with decreasing eliminating tendency of the halogen ions in the sequence Br > Cl > F. Zaugg[2] could prove, e.g., that in the case of decarboxylation of 3-halogen-2,2-diphenyl-propionate (Figure 4) proceeding via halogen elimination, a distinct decrease of the reaction rate with diminishing polarizability of the halogen substituent is observed. The increase of k_{cat} values in the sequence Br < Cl < F therefore excludes the halogen elimination as a rate-limiting step. All other reaction steps succeeding the halogen elimination which are identical for all monohalogenated pyruvic acids have to be ruled out as rate-limiting steps as well. On the other hand, the observed k_{cat} values correlate with the magnitude of the inductive effect of the halogen atom (Figure 5).

The small differences in catalytic constants measured for chloro- and bromopyruvate show up also in the rate constants of hydration of the free acids, as Fischer et al.[3] could show. Thus, a step should be responsible for the differentiation of reaction rates, which depends on the carbonyl activity of the substrate. In keeping with the results obtained with 4-substituted PGA, the addition of the carbonyl group to the C-2 atom of the coenzyme should also be the monomolecular rate-limiting step for halogen pyruvic acids. Since that step is independent of the substrate concentration, a further bimolecular substrate binding step must precede it in the catalysis mechanism. The assumption that this substrate binding step involves the carboxyl group of the substrate is supported by the result given in Table 3. Pyruvamide, the carbonyl activity of which can be compared with that of pyruvate, is only a weak noncompetitive inhibitor, whereas anions of suitable size such as chloride inhibit pyruvate decarboxylation by PDC competitively.

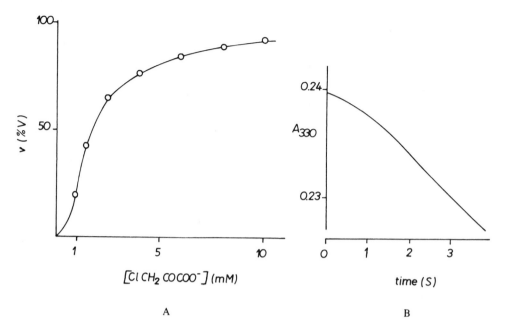

FIGURE 3.   (A) V/S plot of the conversion of chloropyruvic acid by PDC. (B) Progress curve of the conversion of chloropyruvic acid by PDC.

Table 2
KINETIC CONSTANTS FOR THE HALOGEN-ELIMINATING DECARBOXYLATION OF HALOGENPYRUVIC ACIDS BY PDC AND K_i VALUES OF HALOGENPYRUVIC ACIDS (0.1 M CITRATE BUFFER, pH 6.2, T = 25°C)

Substance	k_{cat} (sec^{-1})	K_M (mM)	K_i (mM) (pyruvate as substrate)	Type of inhibition
FCH$_2$ COCOO$^-$	110	2.1		
ClCH$_2$ COCOO$^-$	85	1.9		
BrCH$_2$ COCOO$^-$	80	1.2		
Cl$_2$CH COCOO$^-$	17	—	0.15	Competitive
Br$_2$CH COCOO$^-$	22	0.48		
Cl$_3$C COCOO$^-$	0	—	0.12	Competitive

A further analogy to the oxidative decarboxylation of 4-substituted PGA can be seen in the inactivity of PDC recombined with 4-hydroxy TPP toward halogen pyruvic acids (Table 1). This result again indicates the function of the 4′-amino group within the substrate binding. In Figure 6, a transition state is postulated which is in accordance with the experimental results. Both for the oxidative and the nonoxidative decarboxylation of PGA as well as for the conversion of halogen pyruvic acids proceeding via halogen elimination, this reaction is rate limiting. The mechanism corresponds to an acid-catalyzed carbonyl addition of the substrate to C-2 of the enzyme-bound TPP. Thereby, through an interaction of the carboxylate residue of the substrate with a suitable cationic group of the enzyme, a preorientation of the substrate takes place besides an increase of the carbonyl activity. This preorientation enables the addition reaction of the substrate CO group to the C-2 atom of the enzyme-bound TPP in a monomolecular reaction step.

The protonation of the pyrimidine N1 by a suitable donor group could enhance the acidity

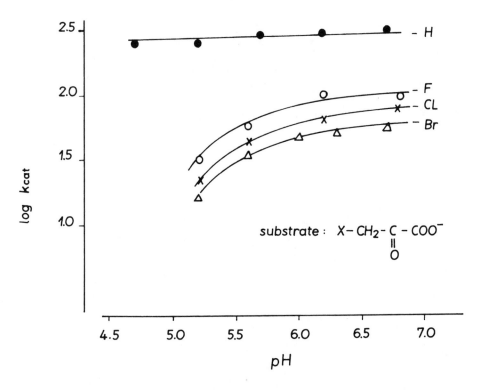

FIGURE 4. Dependence of the catalytic constants for the PDC-catalyzed conversion of pyruvic acid and halogen pyruvic acids on the pH.

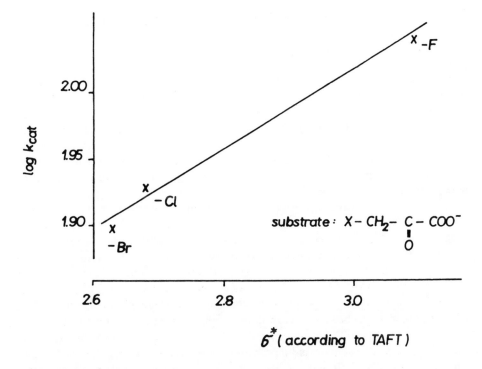

FIGURE 5. Dependence of the catalytic constants for the PDC-catalyzed conversion of halogen pyruvic acids on the halogen atom.

Table 3
INHIBITION OF PDC BY
PYRUVAMIDE AND CHLORIDE
(0.1 *M* CITRATE BUFFER, pH 6.2,
T = 25°C)

Compound	K_i (mM)	Type of inhibition
Pyruvamide	200	Noncompetitive
Cl$^-$	50	Competitive

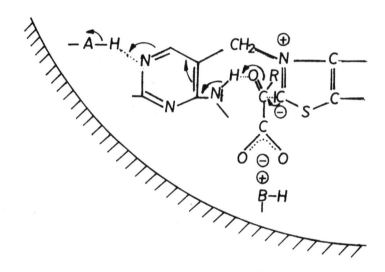

FIGURE 6.   Transition-state structure of the substrate-binding step.

and thereby proton transfer from the 4′-amino group to the carbonyl oxygen. This hypothesis would explain the essential role of N1 during the catalysis mechanism.[4]

Concerning the structure of the transition state for the decarboxylation step, the investigations with halogen pyruvic acids yield several facts. The kinetic constants for the chloropyruvic acids compiled in Table 2 demonstrate that besides monochloropyruvate, dichloro- and dibromopyruvic acid are also metabolized, whereas trichloropyruvate displays no catalytic conversion. On the other hand, a comparison of the inhibitor constants of this type of compound shows that besides the expected inhibiting effect of dichloropyruvate on the pyruvate conversion, trichloropyruvate is also a strong competitive inhibitor. This strong inhibiting effect of trichloropyruvate, not detectable at other basic carbon acid anions which are alike, such as chloroacetic acid, suggests the participation of the carbonyl group in the inhibitor binding within the enzyme-inhibitor complex. Hence, for all three chloropyruvic acids, the substrate binding to the C-2 of the enzyme-bound TPP can be assumed.

While the enzyme-substrate complexes, forming with mono- and dihalogen pyruvic acid, do decarboxylate, the decarboxylation of the enzyme-substrate complex forming with trichloropyruvic acid is blocked. With a calotte model, it can be demonstrated that with metabolizing of dibromopyruvic acid, a covalent bond is only formed between carbonyl-C-atom and C-2 of TPP if the carboxylate group in the forming enzyme-bound active pyruvate is oriented perpendicularly to the plane of the thiazolium ring. This perpendicular fixation of the σ-bond which is to be split toward the electron system is impossible in the adduct arising with trichloropyruvic acid. It appears, therefore, that for decarboxylation the structure of the transition state with perpendicular orientation of the splitting σ-bond toward the thiazolium ring shown in Figure 7 is necessary.

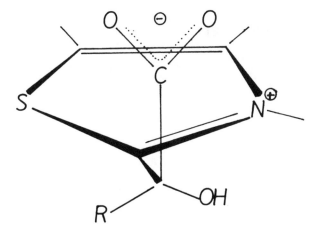

FIGURE 7. Orientation of the splitting σ-bond in the "active pyruvate".

In the step of α-carbanion forming after decarboxylation, the differentiation takes place between native decarboxylation in the presence of H acceptors and the decarboxylation of halogen pyruvic acids proceeding with halogen elimination. In the conversion of 2-oxo acids in the presence of H acceptors such as DCPIP, decarboxylation and oxidative decarboxylation proceed as competitive reactions, whereas in the conversion of halogen pyruvic acids by PDC no chloroacetaldehyde as product of the normal PDC reaction can be detected, even with very sensitive aldehyde tests (e.g., with function sulfurous acid).[5] Obviously, the heterolytic splitting of the C–C bond in the active aldehyde is hampered by the electron-withdrawing halogen atom. There is, it seems, no synchronous decarboxylation-elimination step that would exclude likewise the formation of chloroacetaldehyde as product, since chloropyruvic acid is oxidatively decarboxylated by PDC in the presence of DCPIP (Table 1). Thus, the reaction has to proceed via the step of enzyme-bound active aldehyde and its carbanion, respectively.

Both for the oxidative decarboxylation and for the decarboxylation proceeding via halogen elimination, the last step in the catalysis mechanism, the hydrolytic splitting of acetate out of the enzyme-bound 2-acetyl TPP, is faster than the addition of the substrate to the C-2 of TPP. Differences between native decarboxylation and the conversion of halogen pyruvic acids also appear in the pH dependence of the k_{cat} values (Figure 4).

The native reaction is pH independent in the pH range investigated from 4.6 to 6.6, whereas the k_{cat} values for the conversion of fluoro-, chloro-, and bromopyruvic acid display a distinct pH dependence, the latter indicating a group essential for catalysis with a pK_a value of 5.5 to 6.0 in the enzyme-substrate complex.

In the final reaction step of the native reaction leading to the formation of acetaldehyde, the heterolytic splitting of the C_α–C2 bond occurs. There are two experimental results showing that within the native reaction, this reaction step is partially rate limiting and corresponds in its mechanism to a base-catalyzed splitting of a carbonyl-addition compound: the secondary D-isotope effect for trideutero pyruvic acid of 1.06 and the inverse isotope effect for 4'-$^{15}NH_2$-TPP of 0.994.

The values for the investigated model reaction of pyruvamide, shown in Table 4, display an isotope effect greater than 1 to be found for the splitting of the carbonyl addition reaction, whereas the addition reaction is characterized by effects smaller than 1. The isotope effect, small compared with the basicity of ^{14}N and ^{15}N, for the enzyme recombined with 4'-$^{15}NH_2$-TPP, suggests a participation of the amino group as proton acceptor in the formation of the

Table 4
ISOTOPE EFFECTS IN PDC REACTION

$$\frac{k_{cat}(CH_3 \, COCOO^-)}{k_{cat}(CD_3 \, COCOO^-)} = 1.06$$

$$\frac{k_{dehydration}(CH_3 \, COCONH_2)}{k_{dehydration}(CD_3 \, COCONH_2)} = 1.03$$

$$\frac{k_{hydration}(CH_3 \, COCONH_2)}{k_{hydration}(CD_3 \, COCONH_2)} = 0.89$$

$$\frac{k_{cat}(4' - {}^{14}NH_2 - TPP)}{k_{cat}(4' - {}^{15}NH_2 - TPP)} = 0.994$$

$${}^{15}NH_4^+ + {}^{14}NH_3 \rightleftarrows {}^{15}NH_3 + {}^{14}NH_4^+ \qquad K = 0.962$$

$$\frac{k_{cat}(CH_3 \, CO^{12}COO^-)}{k_{cat}(CH_3 \, CO^{13}COO^-)} = 1.008$$

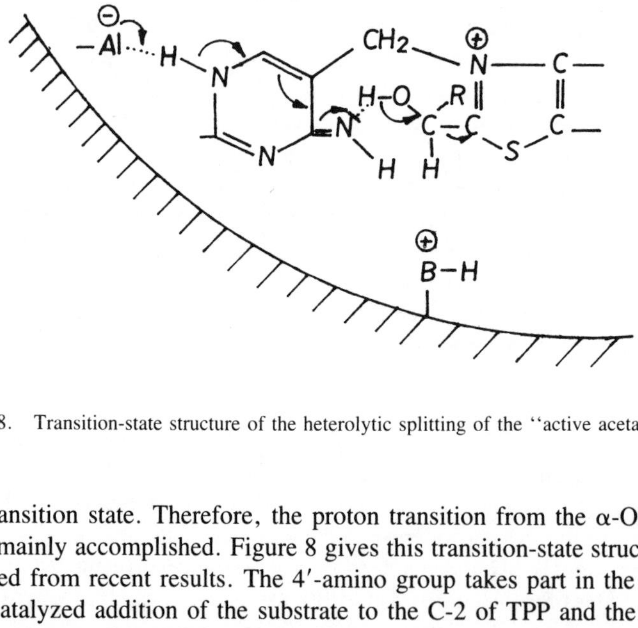

FIGURE 8. Transition-state structure of the heterolytic splitting of the "active acetaldehyde".

rate-limiting transition state. Therefore, the proton transition from the α-OH group to the 4′N of TPP is mainly accomplished. Figure 8 gives this transition-state structure in the way it can be derived from recent results. The 4′-amino group takes part in the decisive steps, i.e., the acid-catalyzed addition of the substrate to the C-2 of TPP and the base-catalyzed elimination of aldehyde from TPP, as proton relays. The rate constants of both reactions seem to be of the same order of magnitude, or the particular substrate thereby causes one or the other step to become rate limiting.

REFERENCES

1. **O'Leary, M. H.**, *Biochem. Biophys. Res. Commun.*, 73, 614, 1976.
2. **Zaugg, M. E.**, *J. Am. Chem. Soc.*, 72, 2998, 1950.
3. **Fischer, G., Sieber, M., and Schellenberger, A.**, *Bioorg. Chem.*, 11, 478, 1982.
4. **Schellenberger, A.**, *Angew. Chem.*, 79, 1050, 1967.
5. **Schiff, H.**, *Liebigs Ann.*, 140, 93, 1866.

Chapter 11

C-13, SECONDARY DEUTERIUM, AND SOLVENT ISOTOPE EFFECTS IN THE ACTION OF PDC

Francisco J. Alvarez and Richard L. Schowen

TABLE OF CONTENTS

I. SUBSTRATE ISOTOPE EFFECTS: GENERAL CONSIDERATIONS

A. Why Kinetic Isotope Effects?

Of the complete armamentarium of the mechanisms scientist, kinetic isotope effects are the most effective technique for application in the study of enzyme action.[1-3] Solvent effects, e.g., have only limited utility because of the instability of many enzymes in nonaqueous media. Substituent effects may give complex or misleading results because of direct interactions between substituent and enzyme, which can affect the rate in a manner not related to the electronic interaction between the substituent and the reacting center. Temperature effects may arise not only from enthalpy differences between substrate in reactant and transition states, but also from thermally induced changes in enzyme structure. Kinetic isotope effects, in contrast, employ only isotopically substituted reactants so that the potential-energy surface (to the level of the Born-Oppenheimer approximation) is unaffected, and the reaction is studied under entirely normal conditions.

Isotopic substitution affects the reaction rate because the vibrational frequencies of the reactants and activated complex are shifted. If the vibrational properties of reactant and transition states are not the same, the shifts will be different in the two states and the resulting difference in vibrational energies and thus free energies of activation will produce an isotope effect on the rate. The isotope effect, usually expressed as a ratio of rate constants (typically k[light]/k[heavy], for the light and heavy isotopic reactants), therefore contains information about the vibrational properties of the transition state and reactant state. Commonly, the reactant state molecule can be studied by vibrational spectroscopy, but the transition state cannot: the kinetic isotope effect is useful just because it contains the same sort of information that vibrational spectroscopy offers for stable states, namely, structural information.

B. Kinetically Complex Reactions

The rate of an enzyme-catalyzed reaction always depends in principle on more than a single rate constant, so that a directly determined isotope effect, e.g., on the initial velocity at a particular substrate concentration, will not refer to a single reactant state and transition state. For useful interpretation, the isotope effects on the individual rate constants must be untangled. Although each of these will refer to the transformation of a single *effective* reactant state (ERS) into a single *effective* transition state (ETS), the common situation is that more than one physical or chemical event contributes to each rate constant, so that it is rarely correct to think of an enzymic isotope effect as illuminating the structure of just one transition state.[4-7]

For example, consider the rate law for the action of pyruvate decarboxylase (PDC) from *Saccharomyces carlbergensis*[8] in which we include a term for substrate inhibition:

$$v = (VS^2)/(A + BS + S^2[1 + (S/K_1)]) \qquad (1)$$

This contains three kinetic terms of mechanistic interest, which are related to the ERS and ETS indicated:

Kinetic term	ERS contributors	ETS contributors
V/A	E + 2S	All transition states after addition of second S to E and before the first irreversible step

Kinetic term	ERS contributors	ETS contributors
V/B	ES + S	Transition states preceding and including decarboxylation of formula ES_2
	E + S	Transition states of formula ES
V	ES	Transition states of formula ES
	ES_2 and subsequent reactant states	Subsequent transition states

The interpretation of the isotope effects on these parameters must obviously take this complexity into account, however, as we hope to show, the complexity itself permits a good deal of useful mechanistic information to be extracted.

C. Determination of Isotope Effects on Kinetic Parameters

The simplest scheme for obtaining the isotope effects for the individual kinetic parameters (e.g., V/A, V/B, and V in Equation 1) is to fit the data for each isotopic substrate to the kinetic law, obtain the parameters, and then the isotope effects as their ratios. This is frequently unsatisfactory, probably because the parameters are statistically correlated, with large estimated errors for the isotope effects. A better scheme, which sometimes allows the determination of isotope effects of a few percent on kinetic parameters that cannot themselves be estimated within hundreds of percent,[9] is the weighting factor method of Stein.[6] For a mechanism composed of sequential steps, like the PDC mechanism, the weighting factor for the i-th kinetic constant, w_i, is given by

$$w_i = v/k(i) \tag{2}$$

where v is the velocity and k(i) is the contribution of the i-th term to the velocity. For example, k(i) = VS/B for the V/B term in Equation 1. Then any isotope effect emerges as a simple weighted average of the isotope effects on the individual kinetic parameters. Letting J(A), J(B), and J(V) signify the isotope effects on V/A, V/B, and V, respectively, we have that

$$v/v' = w_A J(A) + w_B J(B) + w_V J(V), \tag{3}$$

with v and v′ being the two isotopic velocities. The weighting factors in this situation are derived from the velocity and the k(i) for the isotopic substrate whose velocity is in the numerator (unprimed v) of Equation 3. It is therefore necessary to have kinetic parameters for only one of the isotopic modifications in order to compute the weighting factors for the isotope-effect expression. Note that both v/v′ and the weighting factors are functions of substrate concentration S, while the J are substrate independent. Thus, the J can be determined by (1) collecting v/v′ at a series of substrate concentrations; (2) making a kinetic study in some detail for one of the isotopic substrates (naturally, the more convenient) in order to determine the weighting factors; and (3) fitting v/v′ to Equation 3, e.g., by a least-squares procedure, to obtain best-fit values of each of the J.

These remarks apply to the determination of isotope effects by direct rate measurements with two isotopic substrates, each fully labeled with its respective isotope. Only by this method is it in principle possible to obtain isotope effects on each of the kinetic parameters. Another method of isotope-effect determination is the competitive technique, in which a mixture of isotopic substrates is employed. The product is isolated at various times during the reaction, and its isotopic ratio determined and compared with that of the original reactant. From these data, an isotopic velocity ratio is determined. Because the two isotopic species are always present together, competing for the enzyme, it is not possible to determine isotope effects for maximal velocity terms by this method, but only terms of the form V/K.

D. So-Called Intrinsic Isotope Effects

Since a kinetic parameter like V, V/A, or V/B itself commonly refers to a series of steps, the isotope effect will be an averaged values for these steps, with weighting factors playing a role similar to that already discussed for v/v′; again, the slowest step (for V) or the step with the highest free-energy transition state (for V/A or V/B) will contribute the heaviest weight to the observations. The isotope effect for a single elementary event is often called the "intrinsic" isotope effect for that event,[10] although it is probably insufficiently recognized how difficult it may be to isolate a truly elementary event in such a complex dynamical system as an enzyme-catalyzed reaction. Nevertheless, this is occasionally achieved to some degree, and it is from such isotope effects that structural information about the relevant transition states may be extracted. The method generally used is to estimate the expected isotope effect for each possible structure of the transition state, to compare these estimates with the observations, and thus to select that set of structures consistent with the experiment. The estimates are commonly made by either a systematic vibrational-analysis technique[11] or analogy with known isotope effects.

E. Estimates of Expected Isotope Effects by Vibrational Analysis

These estimates are made by systematic exploration of possible transition state structures, construction of a force field for each structure by estimation of force constants for each bond stretch, bend, etc., and calculation of the vibration frequencies for each isotopic modification. A similar calculation is made for the reactant state and, from the frequencies and the Bigeleisen equation,[12] the isotope effect can be calculated. For simple cases, graphical presentations of the results are possible (isotopic maps). For example, contours of constant predicted isotope effect can be plotted against the bond orders of forming and breaking bonds in nucleophilic-displacement reactions,[13] or isotope effect can be plotted against the bond order of the forming bond for carbonyl-addition reactions.[14] The utility of such predictions is limited mainly by the reliability with which transition-state force constants and structures can be postulated, but theoretical calculations of transition state properties by quantum-mechanical techniques are beginning to shed some light on the subject.[15]

F. Estimation of Expected Isotope Effects by Analogy

Vibrational-analysis methods provide a range of estimates of expected isotope effects, corresponding to all possible structures of the transition state. If an intrinsic effect can be determined, it can be compared with these estimates to learn about the structure of the transition state, but in general one has no *a priori* information about what this structure should be. The best one can do if an estimate is needed of the actual intrinsic effect when it cannot be measured directly is to examine the effects in analogs of the step, which are susceptible of isolated study. The risks of this approach are obvious, but it is not without utility, when applied with suitable caution.

II. SUBSTRATE ISOTOPE EFFECTS ON PDC ACTION

A. Kinetic Model and Isotope Effect Determinations

Isotope effects have been determined by direct rate measurement for the following isotopic pairs of substrates, and the isotope effects will be referred to hereafter by the indicated designations:

$$
\begin{array}{lll}
\text{CH}_3\text{CO}^{13}\text{COOH} \text{ vs. } \text{CH}_3\text{COCOOH} & & \text{1-C-13} \\
\text{CH}_3{}^{13}\text{COCOOH} \text{ vs. } \text{CH}_3\text{COCOOH} & & \text{2-C-13} \\
\text{CD}_3\text{COCOOH} \text{ vs. } \text{CH}_3\text{COCOOH} & & \text{3D} \\
\text{CH}_3\text{CD}_2\text{COCOOH} & \text{vs.} & \text{2D} \\
\text{CH}_3\text{CH}_2\text{COCOOH} & &
\end{array}
$$

In all cases, the isotope effects are expressed as k[light]/k[heavy]. The measured velocity ratios, v/v', were fitted to a weighting factor expression based on the kinetic law of Equation 1 to obtain isotope effects on V/A, V/B, and V, with the isotope effect on the inhibition constant constrained at unity.

We will argue that the data are most consistent with a model involving the following sequence of steps, the model being generally consistent also with previous observations by others:[20]

1. Reversible combination of the free enzyme E with substrate S, which binds at a regulatory site
2. A reversible event in which the ES complex just formed is transformed to an active complex for further binding of S
3. Binding of the second molecule of S in the active site
4. Decarboxylation of the adduct of S and thiamin pyrophosphate (TPP), i.e., lactyl TPP or LTTP, to form the ylid or enamine structure
5. Protonation of the ylid/enamine to produce hydroxyethyl TPP or HETPP
6. Decomposition of the HETPP to generate acetaldehyde (or propionaldehyde) and TPP

B. 1-C-13 Isotope Effects

The 1-C-13 isotope effects obtained for the three kinetic terms are

$$
\begin{array}{ll}
\text{V/A} & 1.008 \ (\text{SD } 0.010) \\
& \text{or} \\
& 1.013 \ (\text{SD } 0.005) \\
\text{V/B} & 1.013 \ (\text{SD } 0.024) \\
& \text{or} \\
& (1.000) \\
\text{V} & 1.024 \ (\text{SD } 0.006) \\
& \text{or} \\
& 1.026 \ (\text{SD } 0.003)
\end{array}
$$

The first value cited emerges from an unconstrained fit of the data; isotope effects are then not well determined for V/A and especially for V/B. This is because the V/B term is never highly significant kinetically. Therefore, a second fit was obtained, with the value of the 1-C-13 effect for V/B fixed at unity (second values cited above). This fit is necessarily worse than the first, but negligibly so: in the first case, 63% of the variation is accounted for by the model, in the second case 60%. The constrained fit gives more precise values for V/A and V, with no real change in the estimate for V. The value for V/A may be compared with competitive isotope effects of around 1.008 obtained by O'Leary,[17] Epstein and DeNiro,[18] and Jordan et al.,[19] all of whom isolated the carbon dioxide product and determined its isotopic composition by mass spectrometry.

We begin the interpretation with the effect on V, and we estimate the expected magnitude of the intrinsic effect for the decarboxylation step by analogy. Jordan et al.[19] found that the 1-C-13 isotope effect for nonenzymic decarboxylation of LTPP was about 1.05. This also agrees with a number of other effects of 1.05 to 1.06 for nonenzymic and enzymic decarboxylation collected by O'Leary.[20] We make the mental leap of assuming this enzymic effect to be the same. Assuming that no other step than decarboxylation will generate a 1-C-13 effect, we take: (weight for decarboxylation) (1.05) + (weight for other steps) (1.00) = 1.024. Since the weights must sum to unity, this gives a weight for the decarboxylation step of about 50%. The other steps that can contribute to V are the activation step of ES, the protonation of the ylid/enamine, and the decomposition of HETPP with release of product. The total contribution of these must also be around 50%, but we cannot, until further data are considered, choose among them.

The value of 1.013 for V/A also implies a contribution of decarboxylation to this term. The values of 1.008 obtained by others[17-19] by competitive methods are a good deal more precise than ours, and refer roughly to V/A; furthermore, 1.008 is just at the limit of one standard deviation of our second fit. We consider it prudent, therefore, to take a value for V/A around 1.008. Then the same weighting factor treatment as for V suggests a weight of 8/50 or about 15% for decarboxylation in V/A. The only other step which may contribute to V/A is binding of S into the active site of ES (certainly this "step" may involve several events), and it must therefore be weighted at 85% (being taken to have no 1-C-13 isotope effect).

C. 2-C-13 Isotope Effects

The 2-C-13 isotope effects were determined similarly and subjected to the same two fits (as expected from above, the value for V/B was not well determined):

V/A	1.013 (SD 0.009)
	or
	0.995 (SD 0.007)
V/B	0.951 (SD 0.020)
	or
	(1.000)
V	1.039 (SD 0.004)
	or
	1.032 (SD 0.003)

Here, the second, constrained fit is much less justified than it was above. First, the increased lack of fit is larger (81% of the variation against 90%). Second, there is no chemical justification for neglecting isotope effects at C-2; it is very likely undergoing addition by a sulfhydryl group in the regulatory site.[21] Estimates from the theoretically calculated fractionation factors of Hartshorn and Shiner[22] yield a value of 0.99 for the expected isotope effect. Therefore, we consider the most likely situation to be that the actual isotope effects lie between the limits given above. The validity of a V/A effect in the estimated range is confirmed by the competitive measurements of Epstein and DeNiro,[18] who found values of 1.014 to 1.019. For the sake of argument, we will proceed with mean values of 1.004 (V/A), 0.98 (V/B), and 1.036 (V).

Beginning with V, we require an estimate of the expected 2-C-13 effect for the decarboxylation step. Reasoning by analogy, we consulted a number of measured 2-C-13 effects in nonenzymic reactions[23] and found an average value of about 1.05, in satisfactory agreement with a guess that the isotope effects at the two termini of the breaking bond might be similar. If this is the intrinsic effect for decarboxylation, which has a 50% weight in V, then the other steps with combined weight of 50% must have an average isotope effect of about 1.022, so that the observed value of 1.036 is obtained:

$$(0.5)(1.050) + (0.5)(1.022) = 1.036,$$

The exact value of 1.022 is, of course, unreliable, but the main point is that the effect is sizable, indicating bonding changes at C-2 in the "other steps". These are protonation and decomposition of HETPP, both of which indeed involve bonding changes at C-2.

For V/A, we have already estimated a 15% contribution from decarboxylation. However, in V/A, the effective reactant state is E + 2S, so that the intrinsic effect of 1.05 for decarboxylation must be modified by the equilibrium effects for addition of one molecule of pyruvate to the regulatory site (presumably −SH) and a second molecule of pyruvate to active-site TPP. Taking the Hartshorn-Shiner value of 0.99 for each, we have $(1.05)(0.99)(0.99)$ = 1.03 for the intrinsic overall decarboxylation effect; weighted by 15%, this becomes 1.004. This is in flat agreement with the observation, which then means that the average effect for the binding of the second pyruvate to the active site, which is weighted 85%, must then also be 1.004. Taking once again the equilibrium estimate of 0.99 for the regulatory-site addition, this suggests 1.014 for addition to the active site. This is large enough to suggest that the actual attack of the TPP anion on pyruvate carbonyl may be rate limiting for the binding of the active-site pyruvate.

Of V/B, since it is so ill determined, we will say at this point only that the apparent inverse character (<1) of the effect is consistent with an event of carbonyl addition having preceded the transition state. Among the candidate processes which fill this requirement is conversion of ES to the transition state for activation of the enzyme.

D. 3D and 2D Isotope Effects

These effects are generally rather well determined, being somewhat larger than the carbon isotope effects. The 3D effects are

V/A	0.883 (SD 0.013)
V/B	0.881 (SD 0.026)
V	1.085 (SD 0.006)

If the 2D effects, for 2-oxobutanoate as substrate, were determined by exactly the same mix of reactant states and transition states as the 3D effects for pyruvate as substrate, than the rule of the geometric mean[24] holds that the 2D effects should be given by

$$(3D \text{ effect}) = (2D \text{ effect})^{(3/2)}.$$

The fitted values of the 2D effects and, in brackets, their three-halves powers are

V/A	0.951	(SD	0.012)
	[0.927]		
V/B	0.821	(SD	0.096)
	[0.744]		
V	1.057	(SD	0.005)
	[1.087]		

The values of the 3D and 2D effects for V obviously fit closely the rule of the geometric mean, meaning that use of the "unnatural" substrate does not shift the balance among the rate-limiting steps (assuming they have different isotope effects). There is no obvious agreement in the other two cases but, as we shall see, some reconsideration of the V/B 2D effect will be required.

Much is known about the origin of these effects.[25,26] When a nucleophile, such as the regulatory-site sulfhydryl or the active-site TPP, adds to the carbonyl group of a substrate molecule, electron density is forced out of the pi bond into the adjacent C–H bonds. This strengthens them, producing an isotope effect which, for pyruvate, is 0.88.[25] This should be the limiting, or equilibrium, effect for complete addition and a process which begins with a trigonal carbonyl reactant should exhibit a kinetic isotope effect between 1.00 (transition state still trigonal) and 0.88 (transition state completely tetrahedral). Similarly, a process which begins with a tetrahedral adduct and proceeds toward a trigonal product should produce a normal isotope effect (>1) between 1.00 and $0.88^{-1} = 1.14$.

The 3D effect for V, which arises 50% from decarboxylation and 50% from some combination of ylid/enamine protonation and HETPP decomposition, is quite large and normal. Decarboxylation, formally speaking, involves a change from tetrahedral adduct to trigonal ylid/enamine, but the latter is so electron rich that it is not certain that the adjacent C–H bonds would be weakened; thus, it is possible that no normal isotope effect would result. The large magnitude of the observed average normal effect thus clearly demands that the "other steps" produce substantial normal isotope effects. Since protonation of the ylid/enamine involves a trigonal-to-tetrahedral change and is therefore likely to contribute either an inverse or (as argued earlier) approximately unit effect, the strong indication is that HETPP decomposition constitutes most or all of the "other steps".

If there is no very large contribution of intrinsic 3D effect from the decarboxylation step per se, then the value of 0.88 observed for V/A (which depends to the extent of 15% on decarboxylation) will arise essentially from the binding of the second pyruvate to the active site. The effect is essentially exactly that for a single equilibrium addition (presumably to the regulatory site) and suggests very little nucleophilic interaction of the TPP conjugate base with the second pyruvate at the transition state. If this nucleophile is attacking, the effective transition-state structure must be very reactant-like.

The 3D effect on V/B is the same as the effect on V/A, thus corresponding to one complete carbonyl-addition event ahead of the transition state. The simplest hypothesis is that the activation event is the chief contributor to V/B; if it consists of a conformational change after complete sulfhydryl addition to pyruvate in the regulatory site, then the isotope effect is explained. It is noteworthy that the mean value of the 2D effect on V/B, obtained from an unconstrained fit, is too large in magnitude to be correct: V/B-contributing processes must involve addition to, at most, one pyruvate, with a limiting 2D effect of 0.91. This is just within the large standard deviation for the unconstrained fit. We therefore constrained the V/B effect at 0.91, and obtained values of 0.942 (SD 0.006) for V/A and 1.054 (SD 0.003). These are, in essence, unchanged from the previous values.

III. SOLVENT ISOTOPE EFFECTS: GENERAL CONSIDERATIONS

A. Solvent Isotope Effects and Proton Inventories

In solvent isotope effect studies, the focus is shifted to bonding changes occurring in exchangeable hydrogenic sites of the enzyme and substrate structure, since these sites can be isotopically substituted by immersion of the system in an isotopic solvent. For enzymic investigations, this is, universally, deuterium oxide (DOD).

If an enzymic reaction proceeds more slowly in DOD, it is a sign that exchangeable sites are undergoing a loosening of the binding potential upon formation of the relevant transition state; conversely, if it proceeds more rapidly in DOD, the indication is of a tightening of the mean binding potential for all exchangeable sites. Just as with substrate isotope effects, it is vital to dissect the observations into isotope effects on individual kinetic parameters, so that the interpretation then refers to the binding alterations upon conversion of the effective reactant state for each kinetic term into the effective transition state for that term. Again, these terms will, in general, refer to more than a single elementary event.

A great problem with overall solvent isotope effects k(HOH)/k(DOD) is that they commonly refer to the combined effects from several exchangeable sites. The proton inventory method[27-29] is meant to help in disengaging the combined effects and displaying them individually. It relies on a formulation in which a single parameter, the fractionation factor, is employed to describe the tightness of the binding potential at a given site, compared to the average tightness of binding in bulk water as a standard. A fractionation factor of unity implies a binding potential of equal tightness to that in bulk water; a larger than unit value implies tighter binding and a smaller than unit value implies weaker binding. For the present purposes, the following values are useful to know:

Site	Fractionation factor
Average bulk water molecule	(1.00)
Average protein site	1.0
Hydroxyl group, $-OH$	1.0
Exchangeable site of TPP, $-CH^{23}$	1.0
Sulfhydryl group, $-SH$	0.40—0.46

In the proton inventory method, rates are measured not merely in HOH and DOD, but in mixtures of the two (atom fraction of D = n). Each of the exchangeable sites in reactant and transition states will contribute to the observed change in rate as n is varied. Its contribution is described by a factor:

$$\text{Contribution} = 1 - n + n(\text{fractionation factor}).$$

The rate ratio v(n)/v(HOH) is then given by

$$v(n)/v(HOH) = TSC(n)/RSC(n), \qquad (4)$$

where TSC(n) = product of all transition-state contributions, and RSC(n) = product of all reactant-state contributions. These considerations apply to the proton-inventory expression for a single elementary step; for a kinetic term in which more than one step contributes, an expression like that of Equation 4 must be introduced, with a weighting factor, for each contributing step.

In practice, the method consists of collection of v(n)/v(HOH) as a function of substrate concentration; resolution into the individual k(n)/k(HOH), where k is a kinetic parameter; and analysis of these functions in terms of models for the reactant and transition states (i.e., their fractionation factors). Three simple situations are of particular interest here:

Situation	Manifestation
Isotope effect from single transition-state site	k(n)/k(HOH) linear in n
Isotope effect from single reactant-state site	k(HOH)/k(n) linear in n
Isotope effect from many sites in either state	log {k(n)/k(HOH)} linear in n

These considerations permit the interpretation of our proton-inventory data for the action of PDC.

B. Application of the Proton-Inventory Method to PDC

Measurements of v(n) were made for HOH (n = 0), DOD (n = 0.99), and mixtures between these limits. A weighting factor method was used to resolve the contributions from the individual kinetic terms V/A, V/B, and V. The overall solvent isotope effects, k(HOH)/k(DOD), are

$$
\begin{array}{ll}
\text{V/A} & 0.53 \ (\text{SD} \ 0.02) \\
\text{V/B} & 0.41 \ (\text{SD} \ 0.02) \\
\text{V} & 1.53 \ (\text{SD} \ 0.06)
\end{array}
$$

Thus, on passage from the reactant state to the transition state for the V/A and V/B terms, there is an increase in binding at exchangeable centers, corresponding to the observed inverse solvent isotope effects. For the V term, there is a loosening of binding and thus a normal solvent isotope effect.

A reasonable interpretation for the V/A and V/B terms is suggested by the hypotheses already formulated above, that V/A is determined to the extent of 85% by the combination of S with the activated form of ES, and that V/B is wholly, or nearly wholly, determined by the activation of ES. In both cases, addition of the sulfhydryl group of the regulatory site to the carbonyl group of pyruvate will be complete before the transition state is attained. If this should be the only contributing factor to the solvent isotope effect, then we have:

The rate ratio v(n)/v(HOH) is then given by:

Reactant state: −SH	$RSC(n) = 1 - n + n[0.40]$
Transition state: −OH (i.e., hydroxyl of the adduct)	$TSC(n) = 1 - n + n[1.0] = 1.0$

$$k(n)/k(HOH) = 1.0/(1 - n + n[0.40 - 0.46]$$

$$k(HOH)/k(DOD) = k(HOH)/k(n = 1.0) = 0.40 - 0.46.$$

The prediction is thus that the overall isotope effect will be 0.40 to 0.46; this is in excellent agreement with the effect for V/B, and in satisfactory agreement with the effect of 0.52 for V/A, since the range of 0.40 to 0.46 comes from model experiments. We cannot be sure of the precise fractionation factors in the protein context.

Furthermore, the proton-inventory method gives us a test for the hypothesis; the prediction is that k(HOH)/k(n) should be linear in n in both cases. In fact, this is observed, the correlation coefficient of the linear dependence being 0.999 for V/A and 0.992 for V/B. This is, therefore, strong evidence that essentially the entire solvent isotope effect for these terms comes from sulfhydryl addition at the regulatory site. For the V/A term, some indications were cited above that attack of the TPP carbanion on pyruvate carbonyl might be a contributor. Since the TPP C−H fractionation factor is 1.0, transfer of the proton to most other bases will not produce an equilibrium isotope effect, so there is nothing against the suggestion that the carbanion has been formed and is attacking the substrate. For the V/B term, the hypothesis of the ES unimolecular activation step as chiefly rate limiting holds well.

For the V term, the situation is obviously different: the isotope effect is normal. Furthermore, a plot of k(n)/k(HOH) against n is "superexponential": the curve is bowl-shaped and lies well below the simple exponential curve. In fact, of course, we expect a complex picture here because our previous considerations have indicated that the rate is determined about 50% by decarboxylation and about 50% by a subsequent step, probably the decomposition of the HETPP intermediate or possibly the protonation of the ylid/enamine. The most reasonable guess would be that the decarboxylation step does not generate a solvent isotope effect, and that the observed effect thus comes wholly from one or both of the other two steps. The predicted proton inventory is

$$k(HOH)/k(n) = 0.5 + 0.5\{RSC(n)/TSC(n)\}. \tag{5}$$

Several models which were tested for the V term, along with the percent of the total variation in the data which was fitted by the model, are

Linear least squares	89%
Exponential function	75%
Equation 5, $RSC(n) = 1.0$, $TSC(n) = (1/1.92)^n$	67%
Equation 5, $RSC(n) = 1 - n + n[0.43]$, $TSC(n) = (1/4.47)^n$	92%
Equation 5, $RSC(n) = (1 - n + n[0.43])^2$, $TSC(n) = (1/10.4)^n$	99%

The best fit model, along this line of reasoning, for the nondecarboxylation contribution to V is a rather surprising one: the reactant state is postulated to contain two sulfhydryl groups which participate in the transition state in some sort of proton-transfer event, involving a number of sites and generating altogether a large normal isotope effect of more than 10. It is probably too early to speculate in detail on what the events in question might be (protonation of the ylid/enamine through a chain of hydrogen bonds; general catalysis of HETPP decomposition along a chain of hydrogen bonds), or whether alternative models may also fit the data. For the moment, the result may be taken to illustrate the utility of a proton-inventory investigation of what otherwise seems a rather modest and unremarkable normal solvent isotope effect of around 1.5.

IV. CONCLUDING REMARKS

The final picture which now emerges for the mechanism of PDC action is one in which the decarboxylation event governs the maximal velocity to the extent of around 50%. Some subsequent processes, with the secondary deuterium isotope effects favoring the decomposition of HETPP to products, which may involve two sulfhydryl groups and a multiproton-transfer event, account for the other 50%. Earlier stages of the reaction are indicated to involve a rapid addition of the activator pyruvate to the regulatory sulfhydryl function, a unimolecular activation event following this addition, and then addition of the second pyruvate to the active site. The attack of the TPP carbanion on pyruvate carbonyl may determine the rate for this second pyruvate addition, but only if the transition state occurs early along the reaction path.

ACKNOWLEDGMENT

This research was supported by research grant no. GM-20198 from the National Institute of General Medical Sciences, U.S.

REFERENCES

1. **Melander, L. and Saunders, W. H., Jr.**, *Reaction Rates of Isotopic Molecules*, Wiley Interscience, New York, 1980.
2. **Cleland, W. W., O'Leary, M. H., and Northrop, D. B.**, *Isotope Effects on Enzyme-Catalyzed Reactions*, University Park Press, Baltimore, 1977.
3. **Gandour, R. D. and Schowen, R. L., Eds.**, *Transition States of Biochemical Processes*, Plenum Press, New York, 1978.
4. **Schowen, R. L.**, in *Transition States of Biochemical Processes*, Gandour, R. D., and Schowen, R. L., Eds., Plenum Press, New York, 1978.
5. **Northrop, D. B.**, *Biochemistry*, 20, 4056, 1981.
6. **Stein, R. L.**, *J. Org. Chem.*, 46, 3328, 1981.
7. **Ray, W. J., Jr.**, *Biochemistry*, 22, 4625, 1983.
8. **Hübner, G., Weidhase, R., and Schellenberger, A.**, *Eur. J. Biochem.*, 92, 175, 1978.
9. **Stein, R. L., Fujihara, H., Quinn, D. M., Fischer, G., Kuellertz, G., Barth, A., and Schowen, R. L.**, *J. Am. Chem. Soc.*, 106, 1457, 1984.
10. **Northrop, D. B.**, in *Isotope Effects on Enzyme-Catalyzed Reactions*, Cleland, W. W., O'Leary, M. H., and Northrop, D. B., Eds., University Park Press, Baltimore, 1977.
11. **Buddenbaum, W. and Shiner, V. J.**, in *Isotope Effects on Enzyme-Catalyzed Reactions*, Cleland, W. W., O'Leary, M. H., and Northrop, D. B., Eds., University Park Press, Baltimore, 1977.
12. **Bigeleisen, J. and Wolfsberg, M.**, *Adv. Chem. Phys.*, 1, 15, 1958.
13. **Rodgers, J., Femec, D. A., and Schowen, R. L.**, *J. Am. Chem. Soc.*, 104, 3263, 1982.
14. **Hogg, J. L., Rodgers, J., Kovach, I. M., and Schowen, R. L.**, *J. Am. Chem. Soc.*, 102, 79, 1980.
15. **Williams, I. H., Magglora, G. M., and Schowen, R. L.**, *J. Am. Chem. Soc.*, 102, 7831, 1980.
16. **Sable, H. Z. and Gubler, C. J., Eds.**, *Thiamin: Twenty Years of Progress*, New York Academy of Sciences, New York, 1982.
17. **O'Leary, M. H.**, *Biochem. Biophys. Res. Commun.*, 73, 614, 1976.
18. **Epstein, S. and DeNiro, M. J.**, *Science*, 197, 261, 1977.
19. **Jordan, F., Kuo, D. J., and Monse, E. U.**, *J. Am. Chem. Soc.*, 100, 2872, 1978.
20. **O'Leary, M. H.**, in *Transition States of Biochemical Processes*, Gandour, R. D. and Schowen, R. L., Eds., Plenum Press, New York, 1978.
21. **Schellenberger, A.**, *Angew. Chem. Int. Ed. Engl.*, 6, 1024, 1976.
22. **Hartshorn, S. R. and Shiner, V. J.**, *J. Am. Chem. Soc.*, 94, 9002, 1972.
23. **Alvarez, F. J.**, Ph.D. thesis in Chemistry, University of Kansas, Lawrence, 1985.
24. **Bigeleisen, J.**, *J. Chem. Phys.*, 23, 2264, 1955.
25. **Fischer, G., Kuellertz, G., and Schellenberger, A.**, *Tetrahedron*, 32, 1503, 1976.
26. **Hogg, J. L.**, in *Transition States of Biochemical Processes*, Gandour, R. D. and Schowen, R. L., Eds., Plenum Press, New York, 1978.
27. **Schowen, K. B. J.**, in *Transition States of Biochemical Processes*, Gandour, R. D. and Schowen, R. L., Eds., Plenum Press, New York, 1978.
28. **Schowen, K. B. and Schowen, R. L.**, *Methods Enzymol.*, 87C, 551, 1982.
29. **Venkatasubban, K. S. and Schowen, R. L.**, *CRC Crit. Rev. Biochem.*, 17, 1, 1984.

Chapter 12

REACTIVITY AND FUNCTION OF THE SH GROUPS OF PDC

Alfred Schellenberger, Gerhard Hübner, and M. Sieber

TABLE OF CONTENTS

I. INTRODUCTION

Pyruvate decarboxylase (PDC) has proved, in the course of the past 15 years, an excellent tool and model for studies on the mechanism of enzyme regulation, with a special accent on those mechanisms concerning the aspect of homotropic cooperativity or substrate activation. Some of the reasons, which led to this preference, may be summarized here once more:

1. PDC can be prepared in high yield from brewer's yeast with a relatively low expense of working power. The purified enzyme shows $\alpha_2\beta_2$ structure with respect to its electrophoretic pattern after SDS treatment and a molecular weight of 248,000.
2. PDC-catalyzed reactions result in the formation of three products, the concentrations of which can be estimated easily with common equipment and with high accuracy by means of an optical test (aldehyde), titrimetric methods (titration of the OH^- ions formed with pyruvic acid offers, moreover, the possibility to work at constant substrate concentrations!), and manometric (Warburg technique) or radiometric ($^{14}CO_2$) estimation of the CO_2 formed.
3. PDC shows a significant substrate activation behavior, reflected in a sigmoidal V/S characteristic and a Hill coefficient of n = 1.4 to 1.9 (~2; Figure 1A). The corresponding lag phase in the formation of the products (activation phase; Figure 1B) lasts — depending on the substrate (activator) concentrations — between 1 and 40 sec. This enables very precise estimation of the activation kinetics under varied experimental conditions with the result that all rate constants of the complex activation mechanism have been published already.[1]
4. PDC shows a "total" activation pattern; i.e., the enzyme is completely inactive in the absence of the substrate.
5. Essential for the activation behavior of the substrate molecule is only the R–CO part. Pyruvamide with a missing negative charge at the carboxyl group can take over completely the role of the activator substrate molecule. As pyruvamide has proved to be a weak, noncompetitive inhibitor of the PDC reaction (K_i = 200 mM), no competition between substrate and regulator-substrate molecules is observed, indicating that the negative charge plays a central role in the differentiation of the two mechanisms of catalysis and regulation (as pointed out in more detail by Hübner in Chapter 10).

II. FUNCTION OF SH GROUPS IN THE PDC REACTION

It has been known since the first experiments by Melnick and Stern[2] in 1940 that the catalytic activity of PDC (as that of all other TPP enzymes!) is strongly inhibited by all kinds of SH reagents such as heavy metal ions (Hg^{2+}), p-chloro- and p-hydroxymercuribenzoate (HMB), 2,2'-(dithiobis)-dinitrobenzoate (DTNB), and derivatives of maleinimide. In all cases, the substrate protects from inactivation, which led to the conclusion that SH reagents act as active-site-directed reagents.

The following two reasons have induced us to start new investigations on the mechanistic role of thiols in the PDC mechanism:

1. The catalytic mechanism of TPP enzymes indicates after all — if one neglects interactions between cofactors and protein — no demand of SH groups. Therefore, we had to answer the question: is there perhaps any additional SH-dependent step in the mechanism or does there exist perhaps some other mode that SH groups could influence the catalytic course of the PDC reaction?
2. Since we trusted firmly at that time in the dogma that TPP is able by itself to carry

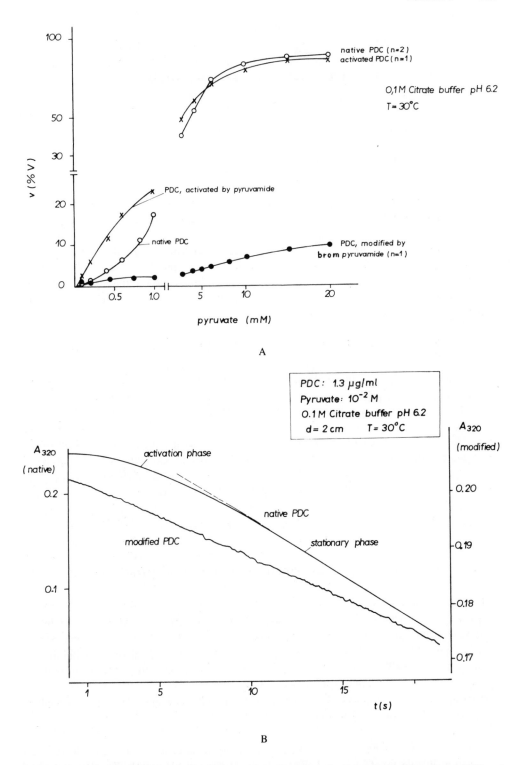

FIGURE 1. (A) V/S plot of native PDC (concentration of pyruvate: 10^{-3} *M*), pyruvamide-activated PDC, and BPA-modified PDC. (B) Time course of activity evolvement of the native PDC and modified PDC. Modifications: preincubation of PDC with pyruvamide (10^{-2} *M*) or incorporation of 6 BPA molecules.

Table 1
NUMBER OF DTNB-REACTIVE SH GROUPS OF
PDC (PHOSPHATE BUFFER pH 7.5) AFTER SDS
DENATURATION (PDC: 248,000 MOL WT)

Enzyme	No. of SH/PDC	No. of S–S/PDC
Holo-PDC	13.87 ± 0.31	
Apo-PDC	13.92 ± 0.26	
PDC treated with 0.05 *M* SDS (pH 9.2)	17.9	2

through the catalytic mechanism, we were practically forced to consider the substrate activation mechanism to be responsible for the SH effects observed.

In the following, we describe, therefore, a series of experiments which were carried out with this motivation, comprising the following steps: (1) number and reactivity of the SH groups of PDC; (2) experiments to demonstrate the specific reaction of the SH reagents with the regulatory sites of the PDC molecule; and (3) proof of a direct participation of SH groups in the substrate activation mechanism of PDC.

III. NUMBER AND REACTIVITY OF THE SH GROUPS OF PDC

Fourteen SH groups can be found per molecule of holo- or apo-PDC after denaturation with 50 m*M* SDS or 6 *M* guanidinium-HCl with DTSB as modification reagent (Table 1). New reference values of the protein component (248,000 mol wt, $A_{280}^{1\%}$ = 11.35 ± 0.10, and extinction coefficient ϵ_{280} = $2.81 \cdot 10^5 \pm 9 \cdot 10^3 \ M^{-1} \ cm^{-1}$ for holo-PDC and $A_{280}^{1\%}$ = 10.52 ± 0.10 for apo-PDC) were taken in these experiments. After reduction of the disulfides by addition of 0.1 *M* dithioerythritol (DTE) to the denatured protein, 18 thiols were found, indicating the presence of two disulfide bridges in the PDC molecule.

The result of a differentiation of the reactivity of the SH groups of native holo-PDC against HMB is shown in Figure 2. The time course of incorporation of HMB, as measured with the help of the stopped-flow technique, exhibits two fractions of SH groups which can be separated by their reactivity: two fast-reacting and four slower-reacting thiols.

In apo-PDC (Figure 3), eight thiols reacting with HMB are observed: two very fast-, four slower-, and two very slow-reacting groups. The number of HMB-reactive SH groups and the pertinent second-order rate constants are summarized in Table 2.

In Figure 4, the decrease of activity of the two-enzyme forms is plotted against the number of HMB molecules incorporated. It is interesting that in holo-PDC, the two fast-reacting thiols can be modified without any loss of activity. The following reaction of the reagent with the four slower-type SH groups reduces the activity of the enzyme to 10% of the native PDC. In the case of the apoenzyme, the catalytic activity of the modified preparations has also been estimated after recombination of the protein with the cofactor. In the case of the holoenzyme, activity drops down to about 10% after the modification of six SH groups. The two very slow-reacting thiols, as well as the modification of other side chains of the protein (no further thiols according to the amino acid analysis), cause a further reduction of the activity of the preparations.

The experiments described show that (1) in the two different proteins, two and three fractions of SH groups exist, showing different reactivity in HMB interaction; (2) TPP protects at least two SH groups from modification (therefore, two different TPP-binding sites should exist in the molecule in accordance to the $\alpha_2\beta_2$ structure); and (3) SH groups are ruled out as binding sites of TPP, because even that apo-PDC which has been modified

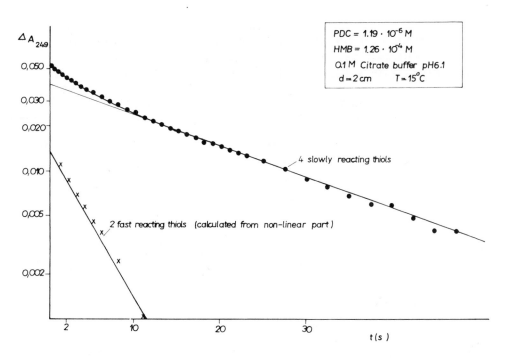

FIGURE 2. Reaction rate of HMB with holo-PDC (conditions: PDC: $1.19 \cdot 10^{-6}$ *M*; HMB: $1.26 \cdot 10^{-4}$ *M*; 0.1 *M* citrate buffer pH 6.1, T = 15°C).

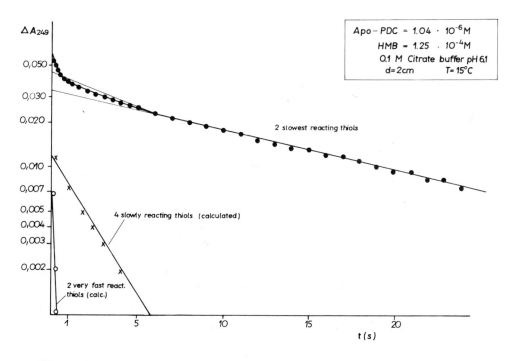

FIGURE 3. Reaction rate of HMB with apo-PDC (conditions: apo-PDC: $1.04 \cdot 10^{-6}$ *M;* HMB: $1.25 \cdot 10^{-4}$ *M;* 0.1 *M* citrate buffer pH 6.1, T = 15°C).

Table 2

**NUMBER AND SECOND-ORDER RATE CONSTANTS FOR THE
MODIFICATION OF THE SH GROUPS OF PDC BY *p*-
HYDROXYMERCURIBENZOATE (CITRATE BUFFER pH 6.1, T: 15°C)**

	No. of reacting SH groups				Rate constants		
Protein	Type 1	Type 2	Type 3	Total	k_1 ($\cdot 10^4$)	k_2 ($\cdot 10^3$)	k_3 ($\cdot 10^2$)
Holo-PDC	—	1.79	3.92	5.71 ± 0.18	—	3.68 ± 0.59	6.56 ± 1.92
Apo-PDC	1.91	1.65	4.09	7.65 ± 0.21	9.13 ± 4.54	6.24 ± 1.52	9.50 ± 1.56

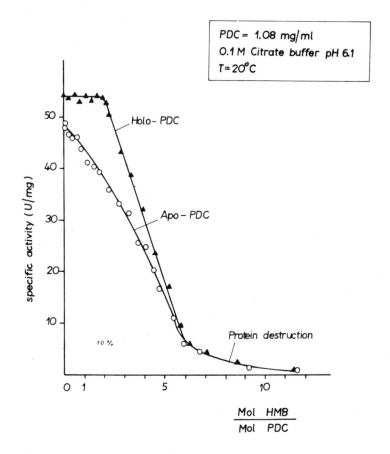

FIGURE 4.   Decrease of the specific activity of holo- and apo-PDC with respect
to the number of HMB molecules incorporated (conditions: PDC: 1.08 mg/mℓ, 0.1
M citrate buffer pH 6.1, T = 20°C).

by eight HMB-reacted SH groups can recombine to the holoenzyme, showing again 10%
residual activity.

IV. SH REAGENTS WITH SPECIFITY FOR REGULATORY SITES

As we have pointed out, pyruvamide takes over completely the regulatory potency of
pyruvate. Experiments were started, therefore, with bromopyruvamide (BPA) as an SH-
modifying reagent with a simultaneous affinity for the regulatory sites. The results obtained
are shown in Figure 5.

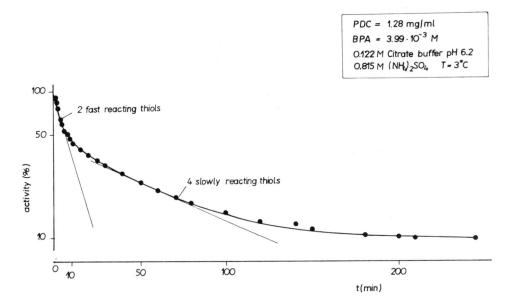

```
PDC = 1.28 mg/ml
BPA = 3.99 · 10⁻³ M
0.122 M Citrate buffer pH 6.2
0.815 M (NH₄)₂SO₄   T = 3°C
```

FIGURE 5.   Decrease of activity (%) of holo-PDC in the presence of a large excess of BPA (conditions: PDC: 1.28 mg/mℓ; BPA: $3.99 \cdot 10^{-3}$ *M*; 0.122 *M* citrate buffer pH 6.2, T = 3°C; (NH₄)₂SO₄: 0.815 *M*).

As in the case of the HMB modification experiments, two types of SH groups are observed in the holo-PDC, and after modification of exactly six thiols, the activity drops down again to 10% of the initial activity. Contrary to the HMB experiments, this value remains constant in the presence of a large excess of the modification reagent as well as after extremely extended reaction times. Consequently, BPA behaves in the case of PDC as a very specific and selective SH reagent.

The enzyme, modified by the incorporation of six BPA and showing 10% residual activity (BPA-PDC) will be characterized in more detail in Section VI.

V. PARTICIPATION OF SH GROUPS IN THE PDC REGULATION?

BPA reacts with PDC via a two-step mechanism.[3] A quick and reversible interaction (probably the formation of a semimercaptal structure) is followed by a slow and quasi-irreversible modification reaction of the thiols (alkylation step). It is not possible, therefore, to reactivate the ultimately formed BPA-PDC complex by addition of DTE (Table 3).

On the other hand, PDC inactivated by HMB can be reactivated completely by the addition of a large excess of DTE. As displayed in Table 3, reactivation is also observed when BPA is added to the reaction mixture *after* treatment of the enzyme with HMB. Consequently, HMB protects the thiols from the "irreversible" modification by BPA, confirming the identical reaction sites of the HMB and BPA interactions.

Let us describe, last but not least, some interesting properties of the BPA-PDC enzyme. As pointed out in Figure 1A, the modified enzyme exhibits a hyperbolic V/S characteristic. Figure 1B confirms this result once more by the missing lag phase in the activity plot. Summarizing, PDC with its covalently modified thiols behaves like a normal Michaelis-Menten-type enzyme.

Table 4 confirms that it is the protein component which causes this modified property. When we prepared from the BPA-PDC enzyme the corresponding apoenzyme and from this (after purification) the holoenzyme, again by addition of the cofactor, an enzyme was obtained showing the same properties as the original BPA-modified PDC: hyperbolic V/S-shape with n = 1 and the full initial activity of 10% with respect to the native PDC.

Table 3
PROOF OF THE IDENTITY OF THE
HYDROXYMERCURIBENZOATE (HMB) AND
BROMOPYRUVAMIDE (BPA)-BINDING SITES OF PDC BY
REACTIVATION OF THE HMB-BLOCKED SH GROUPS WITH
DTE (CITRATE BUFFER pH 6.1)

Experiment	Reactivation with DTE	Time of incubation (min)	Residual activity (%)
Holo-PDC	−	5—420	100
Holo-PDC + HMB	−	10	10
	+	15	98
Holo-PDC + BPA	−	80	10
	+	80	10
Holo-PDC + HMB + BPA[a]	+	48	99

[a]   BPA added after the inactivation of PDC by HMB. Concentrations: PDC: $2.1 \cdot 10^{-6}\ M$;
HMB: $2.31 \cdot 10^{-5}\ M$; BPA: $4.12 \cdot 10^{-3}\ M$; DTE: $10^{-2}\ M$.

Table 4
PROPERTIES OF PDC, MODIFIED BY MEANS OF BROMOPYRUVAMIDE
(SHOWING 10% RESIDUAL ACTIVITY) (BPA-PDC)

Enzyme	Activity (%)	Hill coefficient	K_M (mM)	k^{cat} (sec^{-1})	$\dfrac{\text{Initial rate (t = 0)}}{\text{Steady-state rate}}$
Holo-PDC	100	~2	1.2	316	0
Holo-(BPA-PDC)	10	1	6.0	32	1
Apo-(BPA-PDC)	—	—	—	0	—
Holo-(BPA-PDC) obtained by recombination	10	1		32	1

VI. SUMMARY AND SOME OPEN QUESTIONS

It seems from our experiments that the process of PDC activation depends upon and starts with the interaction of one regulatory substrate molecule with one SH group per regulatory site. Pyruvate acts as regulator species via its carbonyl group, suggesting the formation of a semimercaptal bond, at least as the initiation step of the regulation mechanism.

Moreover, our experiments have pointed out that reversible or irreversible (BPA) modification of the reactive SH groups makes the reactivity of PDC drop down to 10% of the original value. This modified species of PDC shows permanent activity — also in the absence of the substrate — in accordance with hyperbolic V/S plots.

It remains open, of course, which effect of the protein component reduces the rate-limiting step of the PDC reaction to 10% of the original value. It also remains open whether this inactivation mechanism acts on the same process, which causes the activation of the native PDC from 0 to 100% by the substrate. These are only two of the many open questions, but it seems to us that the experiments described offer another possibility to study allosteric processes on a mechanistic basis.

Some additional aspects, confirming the central role of the protein component in the activation mechanism, are presented by König in Chapter 13. We have, nevertheless, the feeling that the "mechanism" of enzyme activation proposed here is only a beginning and actually sounds more like Beethoven's Ninth Symphony blown on a comb.

REFERENCES

1. **Hübner, G., Weidhase, R., and Schellenberger, A.,** *Eur. J. Biochem.,* 92, 175, 1978.
1a. **Schellenberger, A. and Hübner, G.,** *12th Meet. FEBS,* Vol. 52, Pergamon Press, New York, 1979, 331.
2. **Melnick, J. L. and Stern, K. G.,** *Enzymologia,* 8, 129, 1940.
3. **Sieber, M.,** Dissertation, University of East Germany, Halle, 1982.

Chapter 13

KINETIC BEHAVIOR OF PDC CROSS-LINKED WITH BIFUNCTIONAL REAGENTS

Stephan König

TABLE OF CONTENTS

I. INTRODUCTION

In 1970, the research groups of Ullrich and Donner,[9] Boiteux and Hess,[2] and Schellenberger et al.[5] discovered almost at the same time that the enzyme pyruvate decarboxylase (PDC) (EC 4.1.1.1) from brewer's yeast does not belong among enzymes of the Michaelis-Menten type. By extensive measurements at low substrate concentration, a sigmoidal deviation was detected (Figure 1). It is supposed that this kinetic behavior of PDC is caused by a necessary activation of the enzyme by its substrate pyruvate or other artificial activators. This process has to be accompanied by changes of the conformation of the protein. If there exists an activated and inactivated state of PDC derived by conformational changes, it should be possible to freeze them by means of chemical modification using bifunctional cross-linking reagents. After checking several substances, the bisimidoesters were chosen. They possess some advantages favoring them for these purposes. Their solubility in water is excellent so that solvent effects were excluded and they react only with 5-amino groups of lysyl residues. Up to now, no special function has been discovered for these groups.

II. METHODS

The modification procedure was carried out as follows: PDC proved to be pure by sodium dodecyl sulfate (SDS)-polyacrylamide gel electrophoresis (PAGE) was isolated from fresh cells of brewer's yeast according to Sieber et al.[8] and possessed 60 to 80 U/mg protein (1 U is defined as the conversion of 1 μM NADH per minute). PDC (0.2 to 0.4 mg/mℓ) was incubated at least 16 hr at 4°C with bisimidoesters (8 mM/ℓ) of different chain length (0.6 to 1.2 nm;[3] dimethyl glutarimidate, dimethyl adipimidate, and dimethylsuberimidate as hydrochlorides) in various incubation buffers in the absence or presence of pyruvate (100 mM/ℓ). The protein was separated from excess of reagents by ammonium sulfate precipitation and then desalted by gel filtration using Biogel P2 in incubation buffer.

The forms of PDC modified in this way possessed 70 to 80% of the original activity. Their stability in water or aqueous buffer solutions was in the range of that of native PDC or even higher.

Enzyme activity was assayed photometrically by means of indirect optical test according to Holzer et al.[10] with NADH at 340 nm (NADH = 6230 M^{-1} cm^{-1}) and auxiliary enzyme alcohol dehydrogenase on UV/VIS spectral photometer M40 (Zeiss Jena) and, in the case of presteady-state measurements, with stopped-flow equipment (Durrum). Quantitative analysis of the plots of absorbance vs. reaction time was effected with an HP 9825A computer directly connected with the measuring apparatus.

III. RESULTS AND DISCUSSION

I am convinced that the prerequisite of high stability and high catalytic activity are very important for the planned investigations, because it is only possible to get reliable results about conformational changes from kinetic measurements with the help of chemical modification if remaining activity after the treatment is in the range of the original one. This is because loss of activity also means loss of regulatory properties. Previous attempts to freeze activated states of PDC were successful in two cases. The remaining activity in the case of immobilization of PDC at polystyrene beads amounted to 4%[1] and, in the case of modification with 3-halogen pyruvates, only 10% of the original activity.[7]

Although it was possible to distinguish native and cross-linked PDC in the steady state (Figure 2), a residual part of the necessary activation could not be excluded. That is why we decided to measure activation kinetics in the presteady-state phase of the reaction by means of stopped-flow technique. Hübner et al.[6] discovered that the logarithm of the dif-

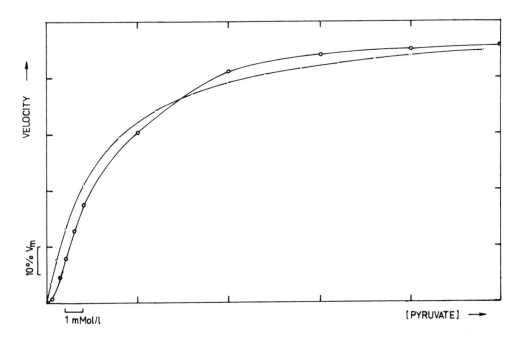

FIGURE 1. Plot of reaction rate vs. pyruvate concentration. Enzymatic activity was assayed at 340 nm with indirect optical test according to Holzer et al.[10] by using 0.2 M sodium citrate buffer pH 6.2. PDC concentration was 6×10^{-9} M/ℓ ($\circ$). ($\cdot$) Indicates a theoretical plot derived from the Michaelis-Menten equation and experimental K_m value. Each data point represents an average of three determinations.

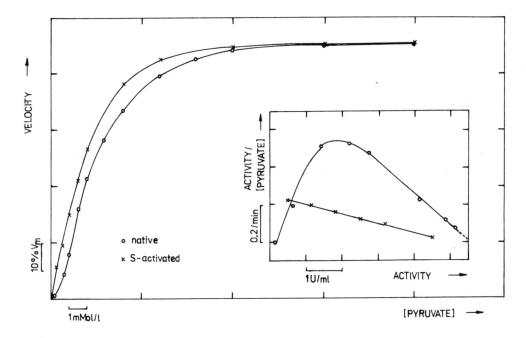

FIGURE 2. Plots of reaction rate vs. pyruvate concentration. Assay conditions are explained in Figure 1. Concentrations of native ($\circ$) and suberimidate-pyruvate-activated PDC (x) were 6×10^{-9} and 1.2×10^{-8} M/ℓ. Insert: EADIE plot of same data points.

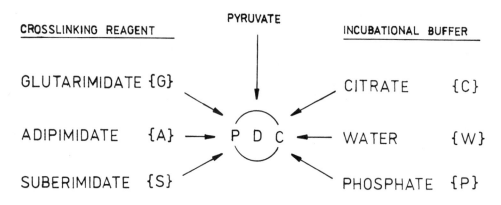

FIGURE 3.   Summary of all 18 modification treatments to which PDC was subjected.

ference between maximal velocity and velocity at time t behaves proportionally to reaction time. Thus, initial activity can be calculated as the initial slope of the reaction curve at its beginning. Figure 3 illustrates a summary of all 18 modification possibilities assayed. The capital letters in brackets indicate the cross-linking reagent used and the buffer in Figures 4 to 7. Citrate is the commonly used buffer for kinetic assays of PDC. Water was applied to prevent chelating of 5-amino groups of lysyl residues of the enzyme by buffer ions. Phosphate is a known inhibitor of PDC, delaying the activation process at relatively high concentrations.[2] Figure 4 reflects a plot of absorbance vs. reaction time of native PDC. The dashed line indicates here, as in all of the other figures, the steady-state rate. The enzyme at the beginning of the reaction is totally inactive; it is activated within 1.5 to 40 sec depending on the substrate concentration.

If the enzyme is modified in the absence of substrate, the plots are similar to those illustrated here (not shown). However, the length of the activation phase was increased to twofold that of the native enzyme. This can be explained by cross-links making conformational changes more difficult. If modification took place in phosphate buffer without pyruvate, the time range necessary to reach the steady-state rate increased to threefold that of native PDC, proving direct influence of phosphate on protein component. However, a totally inactive state of PDC emerged if it was modified with glutarimidate in water. Its activity amounted to 0.3% of the original activity, and no detectable denaturation occurred.

Irreversibly activated forms of the enzyme emerged only after cross-linking with adipimidate and suberimidate. Cross-linking with glutarimidate in the presence of pyruvate led to denaturation of the protein. Figure 5 illustrates the kinetic effect of separating unused activator from the protein as mentioned above. The enzyme states obtained are fully activated from the beginning of the reaction. The initial rate is the steady-state rate. Remaining activity amounted to 70%. In the case of preincubation with auxiliary enzyme, the initial rate decreased to about 50% of the steady-state rate as shown in Figure 6. The same results were obtained if pyruvate was not separated. The time range of activation was as long as that of native PDC. If it was cross-linked in presence of pyruvate and phosphate, no irreversible activation was possible. Kinetic behavior in the presteady-state phase corresponded exactly to that of native enzyme.

Figure 7 indicates a plot of adipimidate-pyruvate-activated PDC after preincubation with auxiliary enzyme and assay in phosphate buffer. Initial activities were dropped down to 20 to 25% that of the steady state. This means that phosphate is able to diminish the extent of activation of modified enzyme.

Summarizing the main facts, I conclude that freezing activated species of PDC by cross-linking with bisimidoesters is possible and gives stable forms of the enzyme. Inactivation

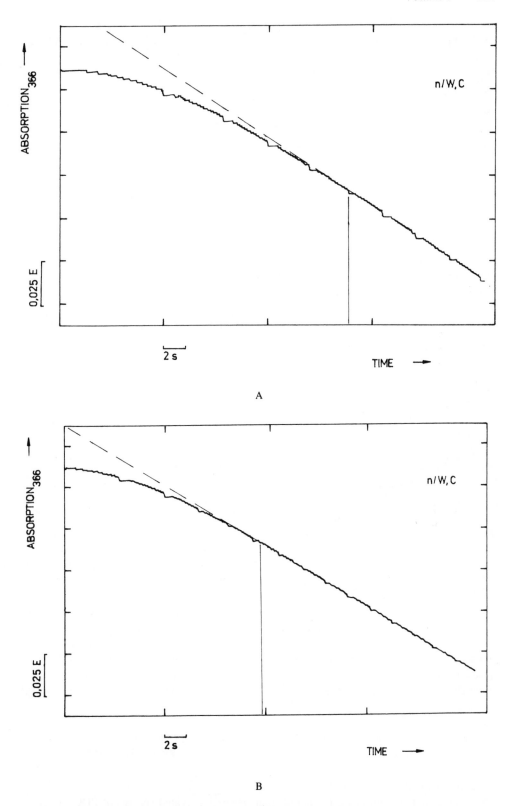

FIGURE 4.   Time course of PDC activation assayed in the presteady-state by means of stopped-flow technique at 30°C in 0.1 *M* sodium citrate buffer pH 6.2 at 366 nm. (A) Pyruvate concentration 5 m*M*/ℓ; (B) pyruvate concentration 25 m*M*/ℓ. Concentration of native PDC was 6×10^{-9} *M*/ℓ.

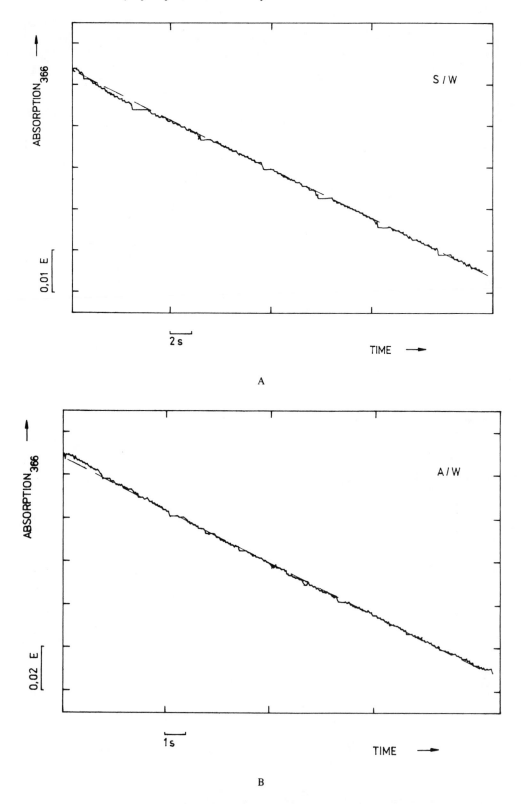

FIGURE 5.   Time course of PDC activation assayed as explained in Figure 4 at substrate concentration of 25 mM/ℓ without preincubation with auxiliary enzyme system. (A) Suberimidate-pyruvate-activated PDC (9×10^{-9} M/ℓ); (B) adipimidate-pyruvate-activated PDC (7×10^{-9} M/ℓ).

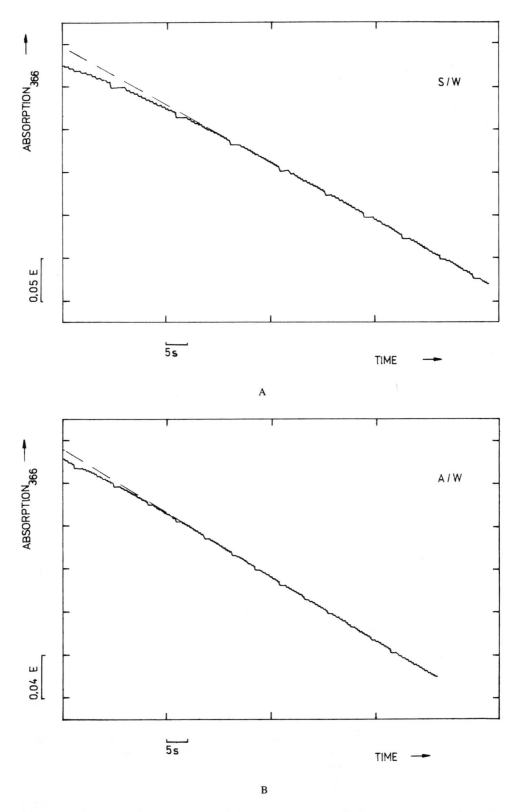

FIGURE 6. Time course of PDC activation after preincubation with auxiliary enzyme system alcohol dehydrogenase/NADH.

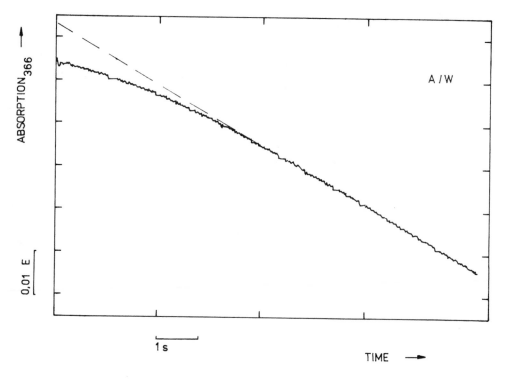

FIGURE 7. Time course of PDC activation after preincubation with auxiliary enzyme system assayed in 0.2 *M* sodium phosphate buffer pH 6.2.

or, better, prevention of activation is successful under special conditions using glutarimidate. Irreversibly activated states arise from modification with adipimidate and suberimidate in the presence of pyruvate. The resulting PDC is totally activated. About 50% of this activation is attributed to bound activator and 50% is attributed to the freezing of the protein conformation. The inhibitor phosphate diminishes the activated state by intensive interaction. However, further investigations are necessary to clarify the influence of this special inhibitor on the protein component of PDC and its conformational changes.

REFERENCES

1. **Beitz, J., Schellenberger, A., Lasch, J., and Fischer, J.,** *Biochim. Biophys. Acta,* 612, 451, 1980.
2. **Boiteux, A. and Hess, B.,** *FEBS Lett.,* 9, 293, 1970.
3. **Hajdu, J., Dombradi, V., Bot, G., and Friedrich, P.,** *Biochemistry,* 18, 4037, 1979.
4. **Hensley, P., Yang, R. Y., and Schachmann, H. K.,** *J. Mol. Biol.,* 152, 131, 1981.
5. **Hübner, G., Fischer, G., and Schellenberger, A.,** *Z. Chem.,* 11, 436, 1970.
6. **Hübner, G., Weidhase, R., and Schellenberger, A.,** *Eur. J. Biochem.,* 92, 175, 1978.
7. **Sieber, M.,** Dissertation, Martin-Luther Universität, Halle-Wittenberg, East Germany, 1981.
8. **Sieber, M., König, S., Hübner, G., and Schellenberger, A.,** *Acta Biomed. Biochim.,* 42, 343, 1983.
9. **Ullrich, J. and Donner, I.,** *Hoppe Seyler's Z. Physiol. Chem.,* 351, 1026, 1970.
10. **Holzer, H., Schultz, G., Villar-Palasi, C., and Jüntgen-Sell, J.,** *Biochem. Z.,* 327, 331, 1956.

Chapter 14

PDH-LIKE REACTION OF PDC

Maria Atanassowa

TABLE OF CONTENTS

I. INTRODUCTION

Pyruvate decarboxylase (PDC) (EC 4.1.1.1) containing thiamin pyrophosphate (TPP) and Mg^{2+} as cofactors catalyzes the decarboxylation of 2-oxoacids to the corresponding aldehydes. Besides this, Holzer and Goedde[1] observed for the first time that in the presence of H acceptors such as 2,6-dichlorphenolindophenol (DCIP), PDC catalyzes the oxidative decarboxylation of 2-oxoacids to the corresponding carbon acids. This reaction, nonphysiological with respect to the enzyme, can be compared with the first step of the oxidative decarboxylation of 2-oxoacids by the pyruvate dehydrogenase (PDH) complex.

We have investigated the kinetics of this reaction with substituted phenyl glyoxylic acids (PGA) as substrate and with analogs of the coenzyme. The experiments should yield insight into the kinetics of this reaction and provide an answer to the question at which intermediate the differentiation between decarboxylation starts to form aldehyde and oxidative decarboxylation in the presence of H acceptors.

Since during oxidative decarboxylation in the presence of H acceptors a quick, so-called paracatalytic, inactivation of PDC[2] takes place, measurements had to be taken by stopped-flow technique to allow a determination of the initial rate.

II. KINETIC MECHANISM OF THE OXIDATIVE REACTION

A typical progress curve for the PDC-catalyzed oxidative decarboxylation of pyruvate by DCIP is shown in Figure 1. Other than the oxidative decarboxylation, nonoxidative decarboxylation occurs as a competitive reaction. In consequence of the conditions mentioned in Figure 1, the oxidative reaction amounts to about 1% of the usual PDC reaction. For the determination of the initial rate, a time interval of up to 2 sec can be used. After longer reaction time, a decrease of the rate occurs due to the paracatalytic inactivation.[2]

The effect of the concentration of the oxidation reagent on the initial rate of the PDC-catalyzed oxidative decarboxylation of pyruvate by DCIP is shown in Figure 2. The hyperbolic dependence of the rate on the concentration of DCIP indicates that the oxidation reagent also forms an enzyme-substrate complex. Furthermore, a nonspecific oxidation of an intermediate formed during enzymic decarboxylation of pyruvate in a bimolecular reaction can be excluded.

In order to determine the mechanism of the two-substrate reaction with DCIP and pyruvate, the dependence of the reaction rate on the concentration of DCIP and pyruvate was investigated. The kinetic investigations have shown that this reaction obeys neither a typical ping-pong mechanism nor a sequential mechanism.

Within a two-substrate reaction, a second reaction of the modified enzyme species (E′) is observed (Figure 3) after the first product (P$_1$) has been split off (competitive ping-pong mechanism). By this reaction, a further product (P$_2$) eliminates, producing the free enzyme, beside the normal reaction with the substrate B. Corresponding to this scheme, the rate equation of the oxidation reaction reads as follows:

$$v = \frac{V_{max} \cdot [A] \cdot [B]}{K_a \cdot [B] + K_B + [A] + [A] \cdot [B] + K_c}$$

[A] — Pyruvate
[B] — 2,6-DCIP   (1)

$$V_{max} = \frac{k_2 \cdot k_5}{k_2 + k_5} \cdot E_o$$ (2)

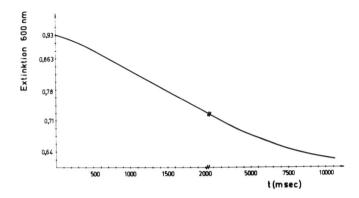

FIGURE 1.   Progress curve of the oxidative decarboxylation of pyruvate by PDC in the presence of DCIP. Pyruvate: 50 mM; DCIP 50 M; PDC: 1.8 M. Path length: 2 cm.

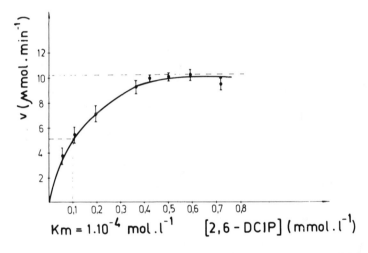

FIGURE 2.   Effect of DCIP concentration on the rate of oxidative decarboxylation of pyruvate by PDC. Pyruvate: 50 mM; PDC: 4.4 M.

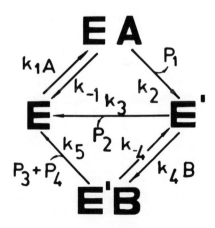

FIGURE 3.   Mechanism of the simultaneous decarboxylation and oxidative decarboxylation of pyruvate by PDG. A: pyruvate; B: DCIP; P_1: CO_2; P_2: acetaldehyde; P_3: acetate; P_4: DCIP H_2; E': enzyme-bound 2-(1-hydroxy-ethyl)-TPP; E'B: Michaelis-Menten complex between E' and DCIP; E: PDC.

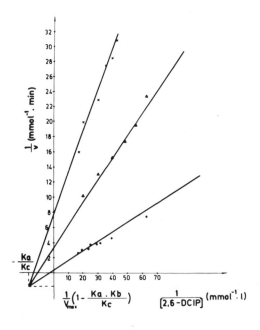

FIGURE 4.   Reciprocal plot for the dependence of the rate of oxidative decarboxylation on the concentration of DCIP and pyruvate. Concentration of pyruvate: (○) $2.5 \cdot 10^{-2}$ M/ℓ; (△) $4.0 \cdot 10^{-3}$ M/ℓ; (x) $2.0 \cdot 10^{-3}$ M/ℓ.

$$K_a = \frac{k_{-1} + k_2}{k_1} \cdot \frac{k_5}{k_2 + k_5} \tag{3}$$

$$K_B = \frac{k_{-4} + k_5}{k_4} + \frac{k_2 + k_3}{k_2 + k_5} \tag{4}$$

$$K_c = \frac{k_2 \cdot k_3 \cdot k_5 + k_{-1} \cdot k_3 \cdot k_{-4} + k_{-1} \cdot k_3 \cdot k_5}{k_1 \cdot k_4 \cdot k_5 + k_1 \cdot k_2 \cdot k_4} \tag{5}$$

where A = concentration of pyruvate; B = concentration of DCIP; K_a = apparent dissociation constant of pyruvate; K_b = apparent dissociation constant of DCIP; and K_c = additive constant.

The reciprocal plot of this equation results in a series of straight lines (Figure 4) which cross each other at an abscissa value of $-K_a/K_c$ for constant concentrations of A and varied concentrations of B. The measured values given in Figure 4 yield $K_a = 30$ mM, $K_b = 0.12$ mM, and $K_c = 2.0$ mM, and are consistent in good approximation with the rate equation derived for the mechanism proposed.

The results correspond particularly to investigations performed with the decarboxylating component of the PDH complex.[3,4] For these investigations, a ping-pong mechanism could be proved with respect to the oxidation of pyruvate by DCIP. Figure 5 demonstrates that inhibition according to the mechanism in Figure 3 is an uncompetitive one since the second substrate (DCIP) reacts with an intermediate (E′) of the PDC reaction.

A. Oxidative Decarboxylation of Substrate Analogs

PGA and its 4-substituted derivatives were proved to be substrates of PDC by Lehmann et al.[5] It was thereby stated that the rate of PDC reaction decreases from 4–OCH₃– to

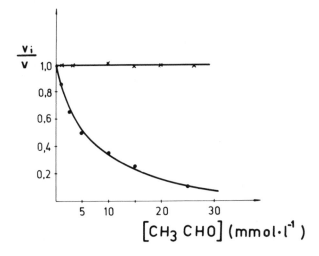

FIGURE 5.   Effect of acetaldehyde on the rate of decarboxylation at 50 m*M* pyruvate (●) and oxidative decarboxylation at 50 m*M* pyruvate and 50 *M* DCIP (x). v: rate in the absence of acetaldehyde; v_i: rate in the presence of acetaldehyde.

4–NO$_2$–PGA. The same acids were also decarboxylated under oxidative conditions, and we have found the same dependence as for the 4-substituent of PGA (see Chapter 10).

III. EFFECT OF ACETALDEHYDE ON THE OXIDATIVE DECARBOXYLATION

The effect of acetaldehyde on the rate of the nonoxidative and oxidative decarboxylation with pyruvate as substrate is shown in Figure 5. The figure demonstrates that in contrast to the aldehyde-forming reaction for which acetaldehyde acts as a noncompetitive inhibitor, the oxidative decarboxylation is not influenced by addition of acetaldehyde. This result indicates that by formation of the enzyme-acetaldehyde complex, only the aldehyde release from the enzyme-bound 2-(1-hydroxy-ethyl)-TPP carbanion (E′) is blocked. All other re-action steps, e.g., the acetate release, are not affected by the binding of acetaldehyde to the enzyme.

IV. EFFECT OF MODIFICATION AT THE COENZYME ON THE OXIDATIVE DECARBOXYLATION OF PYRUVATE

2′-Ethyl-TPP and 4′-hydroxy-4′-desamino-TPP form stable, nondissociating holoenzyme complexes with PDC in the presence of Mg^{2+}. As indicated by the results in Table 1, both the oxidative and the nonoxidative decarboxylation proceed more slowly upon substitution of native TPP by 2′-ethyl-TPP, but the rate of the oxidative decarboxylation is more reduced. This could be due to a greater sensitivity of the oxidative decarboxylation to steric modifications of the cofactor.

A substitution of the 4′-amino group of the coenzyme by an OH group leads generally, for the oxidative reaction as well, to inactivity. This result shows that the 4′-amino group of the coenzyme possesses a function for both substrate binding and aldehyde elimination. It corresponds with investigations carried out with the help of ^{14}C-glyoxylic acid as active-site marker of PDC which indicated that in the 4′-hydroxy-4′-desamino-TPP-holoenzyme complex, the substrate binding is blocked.[6]

Table 1
CATALYTIC CONSTANTS FOR THE DECARBOXYLATION AND OXIDATIVE DECARBOXYLATION OF PYRUVATE BY PDC FOR MODIFIED COENZYMES (0.2 *M* CITRATE BUFFER, pH 6.2, T = 30°C)

Coenzyme	Catalytic constant	
	Decarboxylation	Oxidative decarboxylation
TPP	316	3.8
2'-Ethyl-TPP	152	1.2
4'-Hydroxy-4'-desamino-TPP	0	0

REFERENCES

1. **Holzer, H. and Goedde, H. W.,** *Biochem. Z.,* 327, 192, 1957.
2. **Cogoli-Greuter, M., Hausner, U., and Christen, Ph.,** *Eur. J. Biochem.,* 100, 295, 1979.
3. **Hübner, G., Neef, H., Schellenberger, A., Bernhardt, R., and Khailova, L. S.,** *FEBS Lett.,* 86, 6, 1978.
4. **Tsai, D. S., Hurgett, W. M., and Reed, L. J.,** *J. Biol. Chem.,* 248, 8348, 1976.
5. **Lehmann, G., Fischer, G., Hübner, G., Kohnert, K. -D., and Schellenberger, A.,** *Eur. J. Biochem.,* 32, 83, 1973.
6. **Uhlemann, H.,** Dissertation, Martin Luther University, Halle, East Germany, 1975.

Part III
Transketolase

Chapter 15

STRUCTURAL-FUNCTIONAL RELATIONSHIPS IN BAKER'S YEAST TRANSKETOLASE

G. A. Kochetov

Transketolase (TK) is a two-substrate enzyme which catalyzes the cleavage and formation of ketocompounds by transferring the bicarbon fragment (active glycolaldehyde) from the donor substrate to the acceptor substrate. Baker's yeast TK has a molecular weight of 159,000; it consists of two subunits with an equal molecular weight[1-3] and has two active centers with an equal catalytic activity.[4] The interaction of thiamin pyrophosphate (TPP) with apo-TK is accompanied by typical changes in the optical properties of the enzyme at 255 to 380 nm (Figure 1). Changes at 255 to 300 nm are due to the perturbation of aromatic chromophores in the TK active center. The new wide band at 300 to 380 nm (with a maximum at 320 nm) appears due to a charge transfer complex (CTC) which is formed at the expense of the indolyl group of the apo-TK tryptophan residue (donor) and the TPP thiazolium ring (acceptor).[5,6] The CTC presence accounts for the catalytically active form of the enzyme.[4] Addition of the donor substrate to holo-TK is accompanied by a decrease in the intensity (or even disappearance, depending on the concentration of the substrate added) of the CTC band in the holoenzyme circular dichroism spectrum (CD) and by a negative perturbation in the absorption region of aromatic chromophores (Figure 1).

The overall conformation did not reveal any changes in the TK secondary structure as a result of its interaction with TPP.[7] However, this did not exclude the possibility of delicate conformational changes which did not affect the enzyme secondary structure. To elucidate this matter, we made use of perturbation UV spectrophotometry, a method which makes it possible to evaluate the relative amount of aromatic chromophores contained in the surface layer of the enzyme molecule, their accessibility to the solvent effect, and changes in the chromophore distribution under various conditions.

As it follows from the data given in Figure 2, the perturbation spectrum of apo-TK in 20% dimethyl sulfoxide solution has two positive maxima at 289 and 293 nm.[8] The molar extinction coefficient value for the perturbation spectrum ($\Delta\epsilon_\lambda$) in each maximum is determined by a corresponding contribution of the tyrosine and tryptophan residues accessible to the perturbant.[9] Proceeding from the perturbation data for ethyl esters of *N*-acetyl-L-tryptophan and *N*-acetyl-L-tyrosine,[9] we found that in apo-TK, 4 to 5 tryptophan and 12 to 14 tyrosine residues are accessible to the solvent (Table 1), i.e., about 25 and about 30% of their amount in TK, proceeding from the molecular weight of the enzyme, 159,000.[10] Consequently, most of the aromatic chromophores come to be "immersed" in the interior sphere of the protein molecule. Formation of holo-TK from apo-TK and TPP leads to an abrupt drop in the perturbation spectrum intensity (Figure 2); the amount of the tryptophan and tyrosine residues accessible to the solvent is decreased circa twofold. A subsequent addition of the donor substrate, fructose 6-phosphate (F-6-P), increases the number of accessible tryptophan and tyrosine residues practically back to the initial level (Table 1).

Nothing is known so far about the involvement of tyrosine residues in the formation of the TK active center. As to the tryptophan residues, two are within the TK active centers and form a CTC with the coenzyme.[4] As we have just said, the chemical environment of the two tryptophan residues is changed in the process of holo-TK formation. The reverse process is observed upon subsequent addition of the donor substrate — the two tryptophan residues find themselves in an aqueous environment again. We shall not go into the possible mechanisms responsible for the changes in the environment of the tryptophan residues.

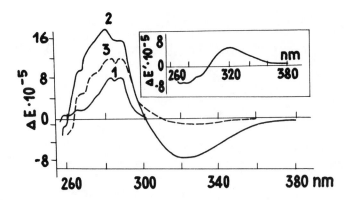

FIGURE 1. CD spectra for apo-TK (1), holo-TK (2), and holo-TK + F-6-P (3). Insert: a differential spectrum obtained by subtracting spectrum 2 from spectrum 3.

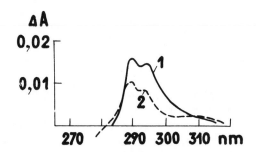

FIGURE 2. TK perturbation differential spectrum in a 20% dimethylsulfoxide solution. (1) apo-TK; (2) holo-TK.

Table 1

$\Delta\epsilon_{289}$ AND $\Delta\epsilon_{293}$ VALUES AND THE AMOUNT OF ACCESSIBLE TRYPTOPHAN AND TYROSINE RESIDUES IN APO- AND HOLO-TK AND THE ENZYME-SUBSTRATE COMPLEX DETERMINED BY PERTURBATION SPECTRA IN A 20% SOLUTION OF DMSO

Enzyme	$\Delta\epsilon_{293}$	Amt. of accessible tryptophan residues	$\Delta\epsilon_{289}$	Amt. of accessible tyrosine residues
Apo-TK	2250 ± 170	4.6 ± 0.4	2540 ± 190	12.9 ± 1.0
Holo-TK	1050 ± 180	2.2 ± 0.4	1410 ± 150	7.6 ± 0.8
Apo-TK + F-6-P	2050	4.2	2600	13.8
Holo-TK + F-6-P	2040 ± 230	4.2 ± 0.2	2830 ± 400	14.5 ± 2.1

However, we can visualize the conformational changes of the apoenzyme, holoenzyme and enzyme-substrate complex this way.[8]

The binding of TPP to apo-TK is accompanied by a redistribution of the surface aromatic chromophores. If there is an equilibrium between forms I and II (Figure 3), TPP shifts the equilibrium toward form II, containing the minimal amount of accessible tyrosine residues (not indicated in the figure) and tryptophan residues, i.e., makes the enzyme more compact in form. The concept about the "loose" conformation of apo-TK and about more "compact"

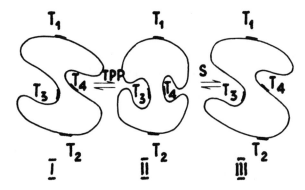

FIGURE 3. T_1- to T_4-TK tryptophan residues accessible to the solvent in apo-TK and the enzyme-substrate complex (forms I and III). T_1- and T_2-tryptophan residues accessible to the solvent in holo-TK (form II).

conformation of holo-TK agrees well with the holoenzyme being more stable than the apoenzyme against heating, changes in the pH of the medium,[7] and trypsin action.[11] It cannot be excluded that the significance of the conformational changes induced by TPP attachment to the protein amounts to the formation of such a structure which ensures holoenzyme stability in an "idle" state and sets the pattern for interaction with the donor substrate. A subsequent attachment of the substrate results in the equilibrium being reshifted toward the open conformation (III). This form has a "loose" and consequently, more mobile surface, thus accounting for a higher probability of contact of the active center functional groups with the solvent and the ligands. Such an enzyme structure can be more receptive to different effects, including regulatory ones.

A comparison of the apo- and holo-TK perturbation spectra does not enable us to say with full certainty which of the tryptophan residues are in the enzyme active center: those which happen to be "hidden" in the holoenzyme, or those which remain "open". However, the fact that all the accessible tryptophan residues are open in both the enzyme-substrate complex and the apoenzyme must attest to the TK active center not being hydrophobic. This is confirmed by our experiments with the use of I-anilinonaphthalene-8-sulfonate as a hydrophobic label and correlates with the previous data.[12]

Thus, the attachment of TPP to apo-TK results in a decrease in the amount of aromatic residues (tryptophan and tyrosine) on the surface of the protein molecule. This process is accompanied by CTC formation between the active center tryptophan residues and TPP. Although the resultant amount of the tryptophan residues inaccessible to the solvent (two residues) coincides with the number of TPP molecules bound to the apoenzyme, there are still no grounds to believe that the active center tryptophanyl residues are "immersed". Therefore, one cannot be certain whether the first step of the catalysis (cleavage of the substrate) proceeds in a hydrophobic or an aqueous medium. In the enzyme-substrate complex, the amount of tryptophan residues accessible to the solvent increases back to the initial value. Consequently, irrespective of the accessibility or inaccessibility of the active center tryptophan residues in the holoenzyme, they become "open" in the enzyme-substrate complex, and so the second step of the catalysis — the transfer of the "active glycolaldehyde" to the acceptor substrate — takes place in the aqueous phase.

Table 2 shows inhibition constants for some TPP analogs. All these analogs operate in a purely competitive mode with respect to TPP.[13] Proceeding from the comparison of the K_i values given in the table and taking into account the higher affinity of TPP for apo-TK than of any other analog investigated, it becomes obvious that the binding of TPP to apo-TK

Table 2
K$_i$ VALUES OF TPP AND ITS ANALOGS^a

Inhibitor	Thiamin	Thiamin monophosphate	Pyrophosphate	Thiazolpyrophosphate
K$_i$(M)	3.4×10^{-2}	2.0×10^{-3}	2.8×10^{-4}	5.4×10^{-6}

^{a} Pyrophosphate, a fragment of the TPP molecule, is designated as a TPP analog, conditionally.

takes place at the very least three points and is effected at the expense of the thiamin part of the coenzyme molecule and two of its phosphate residues. Furthermore, the pyrophosphate residue accounts for the bulk of the contribution to the binding. It is known that a positive charge of the thiazolium ring is necessary for tryptophan-thiamin interaction in the model system.[14] However, a positive charge alone is not sufficient for the enzymatic system: the pyrophosphate group is needed as well (as in the case of TPP), for neither thiamin nor thiamin monophosphate forms a CTC by interacting with apo-TK.[15] This is an essential distinction characterizing the specifics of CTC formation in the enzyme active center compared with the conditions in the model system.

The given data suggest the conclusion that the TPP thiazolium ring, as it forms a complex with the protein, should be oriented accordingly and that this orientation is possible in the case of thiazolium being bound to the pyrophosphate group. The meaning of the orientation effect must be, in all likelihood, to ensure conditions for CTC formation, i.e., the proximity of the donor molecules (the benzene ring of the indolyl group of tryptophan residue) and the acceptor molecules (the positively charged thiazolium ring), and the parallel arrangement of the aromatic rings of both CTC components. The interdependent consecutive steps of the formation of TK active center can be conceived this way (Figure 4).

At the first step, the formation of intermediate complex I takes place. Here, the orienting function of the pyrophosphate residue is not revealed yet, and no CTC can be formed at this stage. The interaction of the pyrophosphate residue with protein limits the number of possible positions for the coenzyme molecule with respect to the contact site. Therefore, there is more likelihood that the TPP thiazolium ring will approach the indolyl group of the tryptophan residue and find itself in a position favorable for interaction (Figure 2).

Yet the role of the pyrophosphate group of the coenzyme may not be limited to the orientation of the TPP thiazolium ring, and here is the reason: the TMP phosphate group participates in binding this analog with apo-TK, as is obvious from the data in Table 2. In other words, it is capable of fulfilling the same anchor function as the TPP pyrophosphate group. Therefore, in line with that stated above about the coenzyme, one could expect a sufficiently high probability of an approach of the positively charged TMP thiazolium ring to the indolyl group of tryptophan residue with favorable orientation. However, according to CD (circular dichroism) data, no CTC is formed in the TMP-apo-TK system. To all appearances, the TMP phosphate residue does not ensure the required mutual orientation of CTC, while in the case of TPP, such orientation becomes possible. The logical supposition is that the pyrophosphate group of the coenzyme is responsible for the orientation of not only the TPP thiazolium ring, but also the tryptophan residue indolyl group in the enzyme active center. CTC formation becomes possible as a result (Figure 4).

Consequently, the TPP pyrophosphate group, interacting with apo-TK, functions not only as an "anchor", but also ensures the orientation of two CTC components in the TK active center — a key condition, in the final analysis, for the formation of a catalytically active molecule of the enzyme.[15]

To identify the other TPP functional groups involved in TPP binding to apo-TK, enzyme analogs were used in which one of the methyl groups was absent: 2'-nor- or 4-nor-TPP.[16]

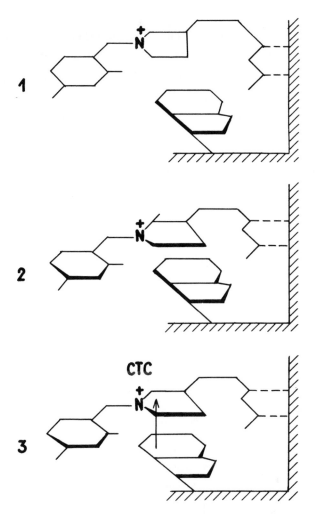

FIGURE 4.   The scheme of TPP interaction with the TK active center.

Dissociation constants were determined by direct titration of apo-TK by the TPP analog with the aid of the CD method.[17] The activity of holo- and apo-TK-TPP analog complexes was determined in a complete system by using both the donor substrate and the acceptor substrate, and also at the intermediate stage of the donor substrate conversion by the α-carbanionic intermediate oxidation rate.[18]

As is apparent from Table 3, the enzymatic activity of both analogs (depending on the TK substrate used) was not exhibited at all, or was significantly lower compared with TPP. The K_d value for 4-nor-TPP proved considerably higher compared with TPP (the K_d for 2'-nor-TPP could not be determined for technical reasons). These data point to the essential significance of both TPP methyl groups for binding the coenzyme to the protein and for a definite TPP conformation determining holo-TK catalytic activity. This is also indicated by the abrupt difference in induced Cotton effects caused by TPP and the analogs (data not shown). These differences may be due to an asymmetry of the optically active chromophores in the active center for TPP and its analogs. All this must testify to a change in the mutual orientation of the protein groups, on the one hand, and separate groups of the coenzyme molecule, on the other, and to overall TPP conformational changes in the TK active center. It can be supposed that the interaction of TPP methyl groups with the protein may stabilize

Table 3
COENZYME ACTIVITY OF TPP ANALOGS IN TK-CATALYZED
OXIDATION AND TRANSFERASE REACTIONS (TPP ACTIVITY IS
TAKEN FOR 100%)

TPP analog	Oxidative reaction involving 1 substrate (donor)		Complete TK reaction involving 2 substrates (donor and acceptor) (activity of TK, %)	
	Substrate	%	Xy-5-P + R-5-P[a]	Hydroxypyruvate + glycolaldehyde[b]
2'-nor-TPP	F-6-P	0		
	Xy-5-P	0		
	DHA	17	0	0
4-nor-TPP	F-6-P	30	7	7
	Xy-5-P	13		
	DHA	0		
4'-N-methyl-TPP	F-6-P	20		
	DHA T	45	0	20
	Hydroxypyruvate	215		
6'-Methyl-TPP	F-6-P	54	50	—
6'-Methyl-4-nor-TPP	F-6-P	0	0	0
	DHA	50		

[a]   R-5-P, ribose 5-phosphate; Xy-5-P, xylulose 5-phosphate.
[b]   The activity assay method is described in Reference 21.

a definite conformation of the coenzyme in the active center — a conformation which determines its catalytic activity.[16]

It is common knowledge that the $C_2=\overset{+}{N}$ group of the TPP thiazolium ring is directly involved in thiamin catalysis. For pyruvate decarboxylase (PDC), a two-center catalysis mechanism has been suggested and substantiated experimentally. In this mechanism, an essential role is assigned, along with the $C_2=\overset{+}{N}$ group, to the 4'-amino group of the coenzyme pyrimidinium ring.[19] One of the proofs of the suggested mechanism is that TPP loses its coenzymic properties in experiments with PDC in the case of any modification of the amino group, and also upon its elimination.[19]

In experiments with TK, only one analog with a modified amino group, 4'-NH-methyl-TPP, was studied in this respect.[20] As seen from the data in Table 3, the effect of amino group modification on the TK reaction rate is significantly dependent at the first step of the reaction on the nature of the substrate used. With F-6-P, the rate is down by 80%, with dihydroxyacetone (DHA) by 55%, and it is even higher with hydroxypyruvate compared to experiments involving an unmodified coenzyme. In all three cases, the α-carbanionic intermediate undergoes oxidation in the given reaction system. This intermediate product is common for all the TK donor substrates. It is clear that at this stage of catalysis (α-carbanionic intermediate oxidation), the nature of the initial substrate does not matter. Hence, the lowering of the TK reaction rate at the first step (the rate is determined by a reaction with ferricyanide), as a result of TPP amino group methylation (Table 3), is due to the retardation of the TK reaction rate at the step preceding the oxidation of the α-carbanionic intermediate — possibly, at the splitting step of the initial substrate. The qualitative assessment poses difficulties for the following reason.

It is known that the rate of a TK-catalyzed oxidation reaction involving the native enzyme and in the presence of one substrate only (the donor substrate) is dozens of times lower than the rate of the TK reaction in a complete system with the two substrates: the donor and acceptor.[18] Consequently, in the former case, α-carbanionic intermediate oxidation (or a release of the reaction product) is a rate-limiting step. At the same time, with 4'-NH-methyl-TPP as a coenzyme, one of the steps preceding α-carbanionic intermediate oxidation becomes rate limiting, since the oxidation rate differs depending on the substrate (see Table 3). It remains to be seen to what extent the rate of this step is lowered as a result of TPP amino group modification; however, it is clear that it decreases much more compared with the overall rate reduction of the oxidation reaction — by 80 and 55%, consequently, with F-6-P and DHA as the substrates (Table 3).

Let us now see how much the TPP amino group modification affects the second step of the TK reaction — the transfer of active glycolaldehyde to the acceptor substrate. Since the enzyme is active in the presence of hydroxypyruvate and glycolaldehyde as the substrates and of 4'-methyl-TPP as the coenzyme (Table 3), the second step of the TK reaction, in principle, is possible with TPP modified at the amino group. At the same time, with Xy-5-P and R-5-P as the substrates, the enzyme does not exhibit any activity with this analog (Table 3).

It should be observed, however, that these experiments did not involve a direct study of the second step of the TK reaction; the aim was to measure the overall rate of the enzymatic reaction in a complete system involving the two substrates, the donor and the acceptor. Therefore, the reduced activity in the former case (hydroxypyruvate + glycolaldehyde) and its absence in the latter case (Xy-5-P + R-5-P) might be due to a drop in the rate of the TK reaction at its first step, different with different substrates, as mentioned above.

Consequently, what we know about the role of the amino group of the coenzyme pyrimidinium ring in the TK reaction is that methylation of the TPP amino group does not prevent TPP from exercising the coenzyme function, but affects the enzymatic reaction rate (at least, at its first step, resulting in the formation of the α-carbanionic intermediate), depending on the substrate used.[20]

According to the two-center mechanism hypothesis based on experiments with PDC, the TPP functional groups participating in the catalysis ($C_2=\overset{+}{N}$ group of the thiazolium ring and the amino group of the pyrimidinium ring) should be oriented accordingly with respect to each other.[19] The TPP conformation in the active center for this orientation was suggested proceeding from the atomic model (Figure 5). In accordance with the model, an incorporation of an additional methyl group at the 6'-position of the pyrimidinium ring should lead to a spatial division of the reaction centers and yield an analog exhibiting no coenzyme activity. However, if in addition to that, the thiazolium ring methyl group is removed as well, no division of the reaction centers is to take place, and the analog should be coenzymatically active.[19] Indeed, such was the case in the studies of these analogs with PDC.

A different picture emerges in experiments involving TK. A TPP analog with an additional methyl group at the pyrimidinium ring 6'-position, inactive in a PDC reaction, is capable of fulfilling the coenzyme function in a TK-catalyzed reaction. Conversely, 6'-methyl-4-nor-TPP, active with PDC, is inactive in a complete TK reaction, though it can bind to the apoenzyme and does not differ in the K_d magnitude from the catalytically active 6'-methyl-TPP.

Consequently, even though the two-center catalysis mechanism may function in the TK reaction, the coenzyme conformation in the TK active center must be different from that in the PDC active center. To elucidate these matters, more studies are needed, including those with the use of other TPP analogs.

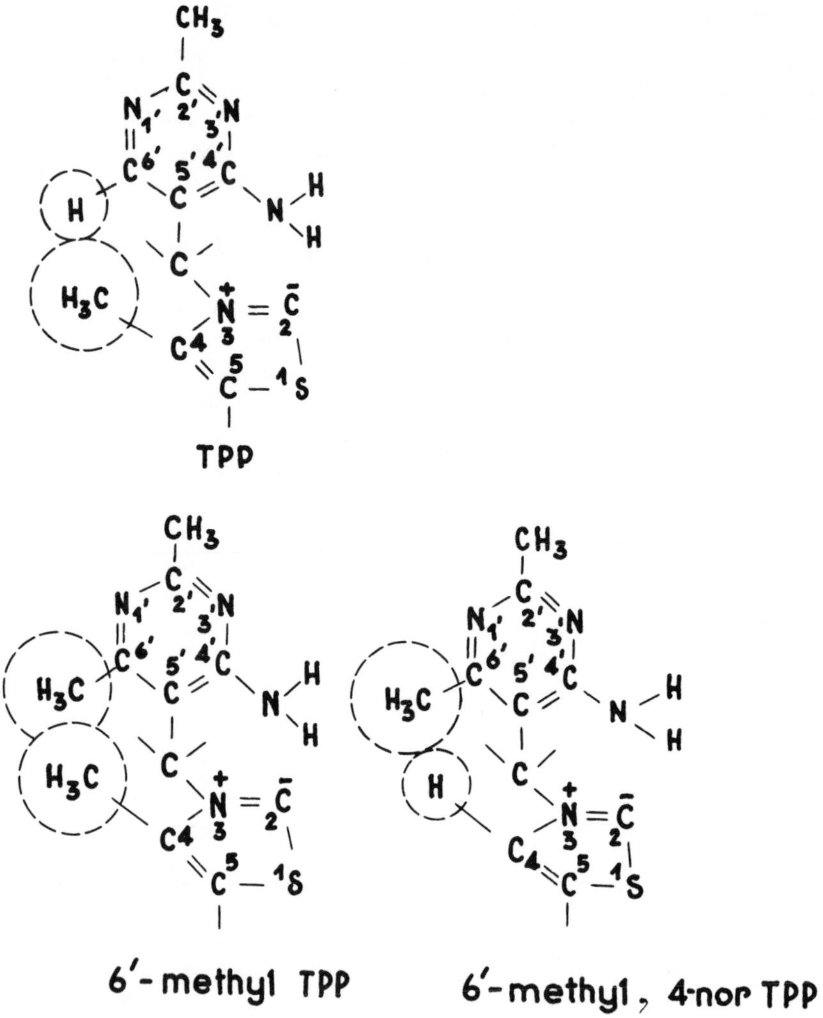

FIGURE 5.   A possible conformation of TPP and its analogs in the PDC active center.

REFERENCES

1. **Cavalieri, S. W., Neet, K. E., and Sable, H. Z.,** *Arch. Biochem. Biophys.,* 171, 527, 1975.
2. **Kochetov, G. A. and Belyaeva, R. H.,** *Biokhimiya,* 37, 233, 1972.
3. **Belyaeva, R. H., Chernyak, V. Ya., Magretova, N. N., and Kochetov, G. A.,** *Biokhimiya,* 43, 545, 1978.
4. **Kochetov, G. A., Meshalkina, L. E., and Usmanov, R. A.,** *Biochem. Biophys. Res. Commun.,* 69, 839, 1976.
5. **Kochetov, G. A. and Usmanov, R. A.,** *Biochem. Biophys. Res. Commun.,* 41, 1134, 1970.
6. **Kochetov, G. A., Usmanov, R. A., and Merzlov, V. P.,** *FEBS Lett.,* 9, 265, 1970.
7. **Kochetov, G. A. and Usmanov, R. A.,** *Biokhimiya,* 35, 611, 1970.
8. **Usmanov, R. A. and Kochetov, G. A.,** *Biokhimiya,* 43, 1796, 1978.
9. **Herskovits, T. T. and Sorensen, M., Sr.,** *Biochemistry,* 7, 2533, 1968.
10. **Kochetov, G. A., Kobylyanskaya, K. R., and Belyanova, L. P.,** *Biokhimiya,* 38, 1303, 1973.
11. **Heinrich, C. P., Moack, K., and Wiss, O.,** *Biochem. Biophys. Res. Commun.,* 49, 1427, 1972.
12. **Heinrich, C. P., Steffen, H., Janser, P., and Wiss, O.,** *Eur. J. Biochem.,* 30, 533, 1972.

13. **Kochetov, G. A., Izotova, A. E., and Meshalkina, L. E.,** *Biochem. Biophys. Res. Commun.,* 43, 1197, 1971.
14. **Biaglow, J. E., Mieyal, J. J., Suchy, J., and Sable, H. Z.,** *J. Biol. Chem.,* 244, 4054, 1969.
15. **Kochetov, G. A. and Usmanov, R. A.,** *Dokl. Akad. Nauk SSSR,* 202, 471, 1972.
16. **Usmanov, R. A., Neef, H., Pustynnikov, M. G., Schellenberger, A., and Kochetov, G. A.,** *Dokl. Akad. Nauk SSSR,* 282, 445, 1985.
17. **Usmanov, R. A. and Kochetov, G. A.,** *Biochem. Int.,* 6, 678, 1963.
18. **Usmanov, R. A. and Kochetov, G. A.,** *Biochem. Int.,* 3, 33, 1981.
19. **Schellenberger, A.,** *Ann. N.Y. Acad. Sci.,* 378, 51, 1982.
20. **Usmanov, R. A., Neef, H., Pustynnikov, M. G., Schellenberger, A., and Kochetov, G. A.,** *Biochem. Int.,* 10, 479, 1985.
21. **Kochetov, G. A., Usmanov, R. A., and Mevkh, A. T.,** *Anal. Biochem.,* 88, 296, 1978.

Chapter 16

PHYSIOLOGICAL FUNCTIONS OF TRANSKETOLASE

Clark J. Gubler

Warburg and co-workers[1a,1b] were the first to recognize a second pathway for glucose metabolism in the oxidation of glucose-6-phosphate (G-6-P) by glucose-6-phosphate dehydrogense (G-6-PDH, then Zwischenferment) using a different coenzyme, coenzyme II, which they called triphosphopyridine nucleotide (TPN, now NADP$^+$).

The various steps of this pathway, known as the pentose phosphate or hexose monophosphate shunt pathway, were subsequently elucidated by the work of Horecker, Racker, and others. It was early thought to be an alternate pathway for the oxidation of G-6-P to produce energy, but it is now generally considered to be a multifunctional metabolic pathway.

Transketolase (TK) was shown to play an integral role in this path by the extensive work of Racker and co-workers, and its role and mechanism of action in this pathway have since been the subject of many studies by groups around the world.

It is well recognized that in tissues, starting with ribose-5-phosphate (R-5-P) as substrate, TK is the limiting enzyme, and this forms the basis for the many modifications of the assay for TK activity available and the widespread use of this assay, performed in the absence and presence (TPP effect) of thiamin diphosphate (ThDP), the coenzyme for TK, for the assessment of thiamin status of animals and man. TK loses its coenzyme more readily than the other ThDP-requiring enzymes, and hence measurement of its activity in blood and tissues, along with the TPP effect, is the most sensitive index of thiamin nutritional status.

Pure homogeneous preparations of TK have been obtained from yeast by Racker and co-workers[2] and more recently by Kochetov and co-workers,[3] and it has been studied in great detail. Most earlier preparations of TK from animal tissues were not pure. Kochetov and co-workers have reported homogeneous preparations from pig[4] and rat[5] livers, but these have not been studied in as great detail.

The generally accepted classical pentose phosphate pathway is shown in Figure 1 (for a more detailed review, see Reference 6). Williams and co-workers[7-9] have recently proposed their L-type pentose phosphate pathway (Figure 2) to explain the finding of arabinose-5-P, sedoheptulose diphosphate, and octulose mono- and diphosphate in liver preparations. This latter scheme widens the scope of the pentose phosphate pathway. In the classical pathway (Figure 1), TK catalyzes two key reactions: the conversion of R-5-P plus xylulose-5-P (Xu-5-P) to glyceraldehyde-3-P (GA-3-P) and sedoheptulose-7-P and Xu-5-P and erythrose-4-P to Ga-3-P and fructose-6-P (Fr-6-P). Transaldolase (TA) is involved in the conversion of GA-3-P and sedoheptulose-7-P to E-4-P and Fr-6-P. Thus, the chief end products are Fr-6-P and GA-3-P which can be recycled through the pathway or are available to the glycolytic pathway, with the action of triose-P and P-hexose isomerases. In their proposed L-type scheme (Figure 2), TK still catalyzes the classical step from R-5-P plus Xu-5-P to GA-3-P and sedoheptulose-7-P, but the second reaction couples E-4-P with D-glycero-D-idooculose-8-P to produce the two hexose phosphates, Fr-6-P + G-6-P, which can then be interconverted by P-hexose isomerase. This scheme does not include TA in the main pathway, but uses aldolase to convert arabinose-5-P (Ar-5-P) and dihydroxyacetone-P to D-glycero-D-idooctulose-1,8-P$_2$ and sedoheptulose-1,7-P$_2$ to dihydroxyacetone-P and E-4-P. This then also requires the action of triose-P-isomerase. Pentose-P-2-isomerase is needed to supply the Ar-5-P from R-5-P. In this case, the end products would be GA-3-P dihydroxyacetone-P and the two hexose-6-phosphates. The true reactions may be a combination of these and may vary with the requirements in the tissue at the time for the various intermediates.

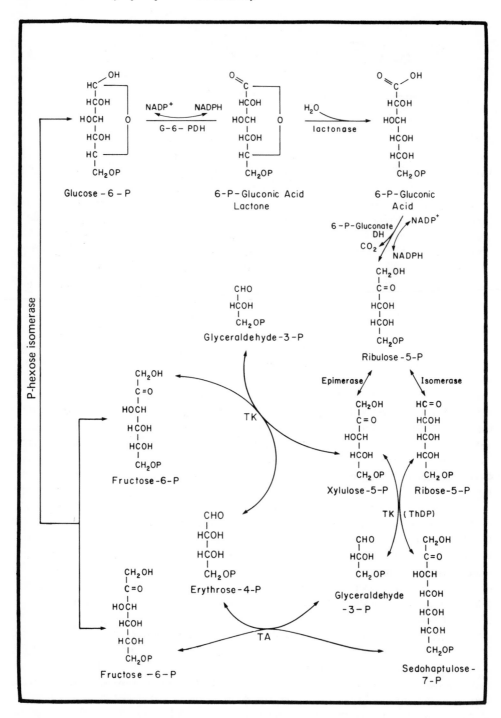

FIGURE 1.   The classical pentose phosphate pathway. (Modified from Horecker, B. L., Paoletti, F., and Williams, J. F., *Ann. N.Y. Acad. Sci.,* 378, 215, 1982.)

The pentose-P pathway occurs primarily in the cytosol along with the glycolytic pathway. It is composed essentially of three systems: (I) an oxidizing or dehydrogenase-decarboxylating system; (II) an isomerizing system; and (III) a sugar rearrangement system. System I oxidizes G-6-P with $NADP^+$ to produce NADPH and CO_2 + Ru-5-P. The $\Delta G^{o\prime}$ of this

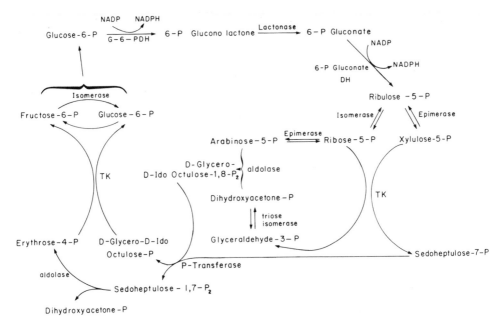

FIGURE 2.   Reactions of L-type pentose phosphate pathway.[7,8]

conversion is -30.8 kJ/mol, which is enough to drive the [NADPH]/[NADP$^+$] ratio to an equilibrium value of over 2000 at 0.05 atm CO_2 pressure. This emphasizes the importance and participation of this pathway in biosynthesis. Catabolic (oxidative) systems utilize NAD$^+$ almost exclusively, whereas both steps in this pathway have an essentially absolute requirement for NADP$^+$. This again suggests a primarily anabolic function rather than for energy production.

The isomerizing system (II) produces an equilibrium mixture of pentose phosphates (Ru-5-P, R-5-P, and Xu-5-P) as required for the various functions. The sugar rearrangment system (III) uses TK and TA in the classical pathway, using two common types of cleavage, and transfer of fragments from a donor to an acceptor: (1) that catalyzed by TK which activates the $-C-C-$ bond next to a $-C=O$ (α-cleavage) to transfer a 2-carbon ketol fragment from a ketose donor to appropriate aldose acceptors, and (2) by TA which activates the $-C-C-$ bond one carbon removed from the $C=O$ (β-cleavage) to transfer a 3-carbon ketol fragment from the donor to an acceptor aldolase. Since these enzymes function with a variety of donor ketoses and acceptor aldoses, this allows considerable versatility to suit the multifunctional needs of the pathway.

In Figure 3, a summary of the various reactions of the pentose phosphate pathway is shown. If these are added, the net reaction seems to be the oxidation of one molecule of G-6-P to CO_2 and H_2O with the production of 12 NADPH. If the NADPH were used for the production of ATP, it would require the NADPH to be used to reduce NAD$^+$, which then could enter the electron transport chain. It is obvious that pentoses, GA-3-P, or other intermediates can be siphoned off at various steps in the pathway without jeopardizing the operation of the pathway. In this case, the net result would, of course, be different and would reflect the needs of the system under those circumstances.

Numerous attempts have been made to measure the contribution of the pentose pathway to the total overall metabolism of G-6-P, by selective ^{14}C-labeling of the G-6-P in the 1, 2, and 6 positions and measuring the amount and position of the ^{14}C-label in the various products. Because of the complexity of this pathway and its close integration with glycolysis, as well as changes to reflect the needs under various conditions, these data are very hard

6-Glucose-6-phosphate + 6 H_2O + 12 NADP → 6-Ribose-5-phosphate + 6 CO_2 + 12 NADPH + 12 H^+

6-Ribose-5-phosphate → 4-Fructose-6-phosphate + 2-Glyceraldehyde-3-phosphate

2-Glyceraldehyde-3-phosphate → Glyceraldehyde-3-phosphate + Dihydroxyacetone phosphate

Glyceraldehyde-3-phosphate + Dihydroxyacetone phosphate → Fructose-1,6-diphosphate

H_2O + Fructose-1,6-diphosphate → Fructose-6-phosphate + Pi

5-Fructose-6-phosphate → 5-Glucose-6-phosphate

Glucose-6-phosphate + 12 NADP + 7 H_2O → 6 CO_2 + 12 NADPH + 12 H + Pi

FIGURE 3.   Summary of net reactions of pentose phosphate pathway.

to interpret quantitatively.[10,11] Generally, the contribution has been found to be relatively small, i.e., usually less than 10%. It undoubtedly plays a larger role in those tissues with a large anabolic requirement, and during periods of increased anabolic activity, i.e., rapid growth, regeneration of tissue, early CNS development, etc.

Whether one considers the "classical" pathway (Figure 1) or the L-type (Figure 2), it is evident that by use of the enzymes normally associated with the pentose pathway or other enzymes available in the cytosol, a large variety of compounds can be produced by this pathway which can be utilized in other systems for other needs. Thus, Fr-6-P, Fr-1, 6-diP, GA-3-P, and dihydroxyacetone-P can be made readily available to the glycolytic pathway. C_3, C_4, C_5, C_6, C_7, and C_8 compounds can be supplied for known and presently unknown metabolic needs. By regulation of the various enzymes, the production of given intermediates can be favored, according to need, without upsetting the system.

Calvin and Basshan[12] have shown that Ru-1, 5-diP is the immediate acceptor for CO_2 in the CO_2 fixation reaction of photosynthesis and have formulated the reductive pentose phosphate pathway known as the Calvin cycle as shown in Figure 4. It can be seen that TK plays a key role in this cycle which accomplishes the fixation of CO_2 to form carbohydrates and the regeneration of the starting material, Ru-5-P. Three ATP are used up for each CO_2 fixed, whereas in the oxidative pentose pathway no ATP is involved.

In experiments designed to determine the biochemical lesion responsible for the marked anorexia associated with thiamin deficiency, rats were made deficient in three ways: (1) thiamin deprivation, (2) oxythiamin (OTh) treatment, and (3) pyrithiamin (PTh) treatment. Then the activities of PyDH and TK were measured in the intestinal mucosa.[13] As shown in Figure 5, anorexia was evident within the first few days of OTh treatment, but did not appear until the 12th to 15th day in thiamin-deficient or PTh-treated rats as evidenced by the growth curves. In Figure 6 it can be seen that the decrease in PyDH is relatively small in all three types of deficiency and does not correlate well with the degree of anorexia. On the other hand, in Figure 7 it is evident that the TK activity decrease is more marked, and is most marked particularly in OTh-treated rats, which also show the earlier and more severe signs of anorexia. The decrease in TK activity follows closely the appearance and degree of anorexia observed. From this it was concluded that TK must play some significant role in the development of the anorexia, i.e., in the function of the mucosa.

In extensive studies on the functions of thiamin in brain function, Dreyfus and Hauser[14] have shown that PyDH activity is only slightly reduced, even in severe thiamin deficiency showing neurological symptoms, whereas TK activity is markedly reduced parallel with the

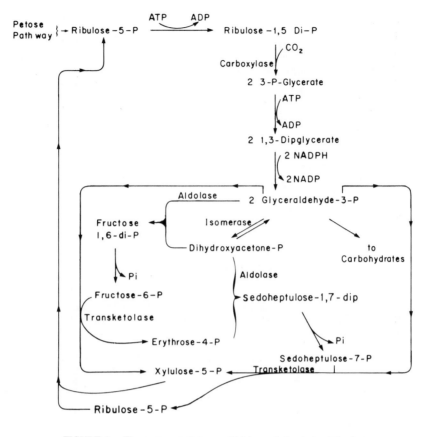

FIGURE 4. The pentose pathway or Calvin cycle in photosynthesis.

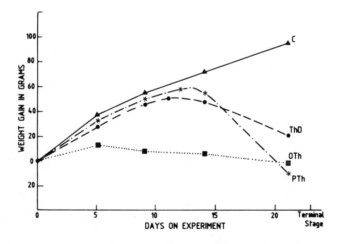

FIGURE 5. Weight change curves of rats in experimental thiamin deficiencies. C = control; ThD = thiamin deprivation; OTh = OTh treated; PTh = PTh treated. (From Bai, P., Bennion, M., and Gubler, C. J., *J. Nutr.*, 101, 731, 1971. With permission.)

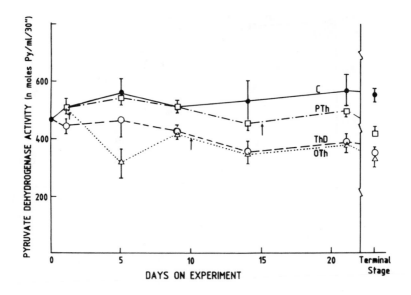

FIGURE 6.   Mucosal pyruvate dehydrogenase activity of rat small intestines. Arrows show the time appearance of anorexia. (From Bai, P., Bennion, M., and Gubler, C. J., *J. Nutr.*, 101, 731, 1971. With permission.)

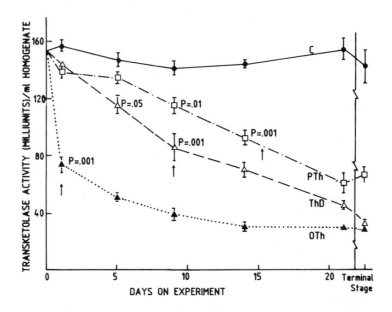

FIGURE 7.   Mucosal TK activity of rat small intestines. Arrows show the time of appearance of anorexia. (From Bai, P., Bennion, M., and Gubler, C. J., *J. Nutr.*, 101, 731, 1971. With permission.)

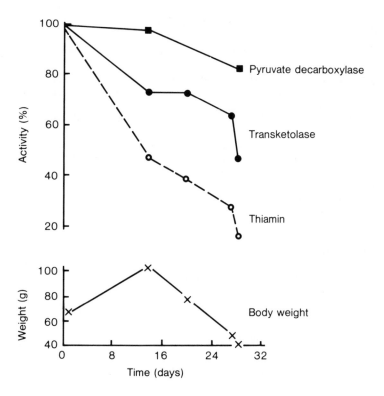

FIGURE 8. Thiamin content, TK, and pyruvate dehydrogenase activity in brain in thiamin deficiency. (From Dreyfus, P. M. and Hauser, G., *Biochim. Biophys. Acta,* 104, 78, 1965. With permission.)

1. $G\text{-}6\text{-}P + 12\,NADP^+ + 7\,H_2O \longrightarrow 6\,CO_2 +$
   $12\,NADPH + 12\,H^+ + Pi$
   6 turns of pathway (minor)
2. Supply reducing power ($NADPH + H^+$) for anabolic processes (biosynthesis):
   (a) of fatty acids
   (b) of cholesterol and steroids
   (c) Reductive carboxylation of pyruvate to malate for reversal of glycolysis and regeneration of 4-C-dicarboxylic acids for citric acid cycle.
3. Supply of ribose-5-P for production of nucleotides and nucleic acids.
4. Supply of Ribulose-1,5-di-P in photosynthetic production of carbohydrate.
5. Supply of glycolytic intermediates, Fr-6-P and GA-3-P.
6. Role in integrity of red cells.

FIGURE 9. Summary of functions of TK.

reduction in thiamin level and the development of the neurological symptoms (Figure 8). They suggest that it is the reduction in TK activity which may be the lesion responsible for initiation of the clinical and histopathological events.[14] It has been demonstrated repeatedly that TK loses its coenzyme more easily than does PyDH and hence is a more sensitive index of early thiamin deficiency. This is particularly true of the red blood cells and is the basis for the use of TK assay for the assessment of thiamin nutritional status.[15]

In Figure 9, the known functions of TK as a key enzyme in the pentose phosphate pathway are summarized. In addition to these functions, it may play a role in maintaining the integrity of the red cells, in myelin production in the central nervous system, especially in the glial cells and in the supply of C_3, C_4, C_5, C_6, C_7, and C_8 intermediates for use in a variety of functions. TK also plays a role in specialized fermentation schemes in some bacterial species.

REFERENCES

1a. **Warburg, O., Christian, W., and Griese, H.,** *Biochem. Z.,* 282, 157, 1935.

1b. **Warburg, O. and Christian, W.,** *Biochem. Z.,* 292, 287, 1937.

2. **DeLaHaba, G., Leder, I. G., and Racker, E.,** *J. Biol. Chem.,* 214, 409, 1955.

3. **Kochetov, G. A., Meshalkina, L. E., and Usmanov, R. A.,** *Biochem. Biophys. Res. Commun.,* 69, 839, 1976.

4. **Philippov, P. P., Shestakova, I. K., Tikhomirova, N. K., and Kochetov, G. A.,** *Biochim. Biophys. Acta,* 613, 359, 1980.

5. **Kochetov, G. A. and Minin, J.,** *Biokhimiya (Moscow),* 43, 631, 1978.

6. **Horecker, B. L.,** in *Reflections on Biochemistry,* Kornberg, A., Horecker, B. L., Cormundella, L., and Oro, J., Eds., Pergamon Press, Oxford, 1976, 65.

7. **Horecker, B. L., Paoletti, F., and Williams, J. F.,** Thiamin — twenty years of progress, *Ann. N.Y. Acad. Sci.,* 378, 215, 1982.

8. **Williams, J. F.,** *TIBS,* 5, 315, 1980.

9. **Williams, J. F., Blackmore, P. F., and Clark, M. G.,** *Biochem. J.,* 176, 257, 1978.

10. **Williams, J. F., Rienits, K. G., Schofield, P. J., and Clark, M. G.,** *Biochem. J.,* 123, 923, 1971.

11. **Wood, H. G., Katz, J., and Landau, B. R.,** *Biochem. J.,* 338, 809, 1963.

12. **Calvin, M. and Basshan, J. A.,** *The Photosynthesis of Carbon Compounds,* Benjamin, New York, 1962.

13. **Bai, P., Bennion, M., and Gubler, C. J.,** *J. Nutr.,* 101, 731, 1971.

14. **Dreyfus, P. M. and Hauser, G.,** *Biochim. Biophys. Acta,* 104, 78, 1965.

15. **Brin, M.,** in *Thiamin Deficiency: Biochemical Lesions and Their Clinical Significance,* Wolstenholme, E. E. W., Ed., Churchill, Edinburgh, 1967, 87.

Index

INDEX